Causes of CLIMATE CHANGE

Ashok Malik

RAJAT
PUBLICATIONS
New Delhi, India

CAUSES OF CLIMATE CHANGE

by *Ashok Malik*

This edition is published by Rajat publications
4740/23, Ansari Road Daryaganj,
New Delhi - 110002

ISBN 978-81-7880-341-8

Edition 2026

Printed and bound in India

RAJAT
PUBLICATIONS

4740/23, Ansari Road, Daryaganj,
New Delhi- 110002 (India)
Tel:- +91-11-23267924
Mob:- +91-9811484973, +91-9811105389
E-mail: rajat_publication@yahoo.com

Preface

Climate change, or global warming to be more precise, is to day an issue no longer confined to the realm of scientific interest and research but one which has spilled onto the arena of public interest and cooperation to combat it the phenomena of the increase in the global mean temperature has wide reaching consequences and until some drastic, long term plans are undertaken to fight these consequences, life as we know today threatens to alter dramatically for the worse, on even case to exist.

The drawn of the fast paced scientific age, with its consequent industrialisation and urbanisation had always troubled enlightened scientists, and global warming was at the helm of pressing concerns troubling them. Repeated warming of these scientists, throughout the past two decades, to slow down this process, had fallen on deaf ears. But now, with unprecedented economic and social growth, not only governments, but the entire world public can no longer afford to turn a blind eye to this issue. Global warming ominously promises to wreak have on the planet, and its consequences are becoming ever more conspicuous. Ecosystem and biodiversity fragmentation, unusual weather conditions, rising sea levels and decrease in food productivity have already shown increasing frequency in their occurrence, and unless something is contrived to fight this menace, these occurrences would show no signs of relenting.

This book, the first in the series of this four book volume on climate change, deals, exclusively with the exploration of the causes of climate change. The increase in the emission of green house gases and the subsequent enhanced greenhouse effect, is the root of this problem, and the human activities and processes which cause these are quite comprehensively discussed. Trends in the increase in the concentrations of these green house gases are detailed with the use of tables and graphs, and pie - charts have also been included to show the contribution of different gases to global warming. The scientific chemical processes to show the production of these gases, also find cover here further illustrations and photographs to ascertain the impacts of these gases a are also included. In addition the consequences of the actions of these causes are discussed and critical insights into combating these causes one listed. To conclude the compendious information on the subject, lucid language and content included in such a way so as to faster clear and insightful understanding of the issue make this book a must read for all those who are interested in knowledge about the roots of global climate change.

Ashok Malik

Contents

1

The Science of Climate Change

A variety of factors determine the rate and magnitude of climate change, including the emissions of greenhouse and aerosol-producing gases, the carbon cycle, the oceans, biosphere, and clouds. As understanding in each of these areas evolves, it is important that researchers, policy-makers, the press, and the public be kept informed since these developments affect of the seriousness and complexity of this issue.

The temperature rise is expected to be greater in the U.S. than the average temperature increase across the globe. While changes in precipitation and extreme weather events such as hurricanes and other storms are more difficult to predict, it is possible that the intensity of rain and hurricane events could increase. Uncertainties in predicting the direction and magnitude of these changes make it difficult to predict the impacts of climate change. However, even small changes in climate can lead to effects that are far from trivial.

Observed Changes in Climate

Observed changes in climate and in the factors that may be responsible for these changes. The main concern is the human influence on climate. Other fctors are also considered, since these form the backdrop against which human influences are imposed.

Atmospheric Composition Changes

The composition of the atmosphere has changed markedly since pre-industrial times: CO_2 concentration has risen from about 270-280 parts per million by volume (ppm) to over 360 ppm today, CH_4 has risen from about 700 parts per billion by volume (ppb) to over 1700 ppb, and N_2O has increased from about 270 ppb to over 310 ppb. Halocarbons that do not exist naturally are now present in substantial amounts. The pre-industrial levels of these gases are known because the composition of ancient air trapped in bubbles in ice cores from Antarctica can be measured directly. These ice cores show that the changes since pre-industrial times far exceed any changes that occurred in the preceding 10,000 years.

Human activities—fossil-fuel burning, land-use changes, agricultural activity, the production and use of halocarbons, etc.—are the dominant cause of these changes. This is undeniable for halocarbons like CFC11 and CFC12 because these gases do not occur naturally. For CO_2, CH_4, and N_2O, the human role is virtually certain too, partly because of the rapidity of changes since preindustrial times, but also because the changes can be well simulated using appropriate models driven by past emissions changes.

For CO_2, analyses of radiocarbon changes prove that emissions from fossil-fuel combustion have been a major contributor to the concentration increase. Land-use changes have also contributed significantly. For CH_4, the primary sources have been agriculture, animal husbandry, land-fill emissions, and leakage associated with fossil-fuel production and distribution. The main source for N_2O appears to be linked to the use of nitrogen compounds in agriculture as fertilisers. For these three gases, their total emissions are reasonably well defined. Their emissions "budgets" are more uncertain. The gases do, of course, have important natural sources. However, in preindustrial times the sources were balanced by natural removal or "sink" processes: by fluxes into the oceans and terrestrial biosphere for CO_2, and, for CH_4 and N2O, mainly by chemical reactions in the atmosphere. Human activities have disturbed these balances.

For the halocarbons, the most climatically important of which are the chloro fluorocarbons CFC11 and CFC12, the sources are almost all anthropogenic. Today, these sources are largely controlled under the Montreal Protocol and its Amendments and Adjustments. However, new "substitute" chemicals, which are not controlled because they do not cause depletion of stratospheric ozone, are being introduced. These new gases, like all halocarbons, are strong greenhouse gases.

In addition to the gases mentioned above, there have been other important atmospheric composition changes due to anthropogenic activities. Emissions of gases like carbon monoxide (CO), nitrogen oxides (NOx), and volatile organic compounds (VOCs) such as butane and propane, which have resulted from industrial activity and land-use changes, haveled to large changes in tropospheric ozone. Tropospheric ozone is a powerful greenhouse gas.

Radiative Forcing Changes

The above changes in atmospheric composition have disturbed the overall energy budget of the planet, upsetting the balance between incoming short-wave radiation and outgoing long-wave radiation—the planet's "radiative balance." Such a change is referred to as "radiative forcing." The climate system responds to positive radiative forcing by trying to restore the radiative balance, which it does by warming the lower atmosphere. The larger the radiative forcing, the larger the eventual surface temperature change.

For each greenhouse gas, and for sulphate and other aerosols, it is possible to calculate the corresponding global-mean radiative forcing. By adding the separate forcings together. Information on the relationships between forcing and concentration changes has been given by the IPCC.

For the greenhouse gases, the individual components may be uncertain by up to ±10 percent. For total greenhouse-gas forcing, the uncertainty is probably similar. For sulphate aerosol forcing the uncertainty is considerably larger than for greenhouse gases, particularly for the indirect aerosol forcing effect. For the relatively long-lived gases, the spatial patterns of radiative forcing are fairly uniform. For short -lived constituents, which have lifetimes of only days to weeks, because their concentration changes are much larger near their sources than elsewhere, the spatial patterns of radiative forcing vary markedly from place to place. Thus, to determine the regional details of past and future climate change. For the other gases it is sufficient to know only their global emissions changes.

The climate system has experienced more than just anthropogenic forcing since pre-industrial times. In addition, there is strong—but indirect—evidence that appreciable changes have occurred in the energy output of the sun, both on the sunspot cycle time scale and on longer time scales. A number of attempts have been made to reconstruct past changes in the sun's output using sunspot and related data, information from other sun-like stars, etc. Prior to the satellite era, even though these reconstructions show qualitatively similar changes, they remain highly uncertain.

Global-Mean Temperature Changes

The simplest and most revealing index of climate change is the global-mean temperature near the Earth's surface. Analysis of this record provides valuable insights into the causes of past climate change. The standard record used by the IPCC combines land data developed in the Climatic Research Unit and marine temperature data compiled by the U.K. Hadley Centre. The raw input data for these records come from many sources, and are subject to numerous inconsistencies arising from nonclimatic effects such as changes in instrumentation, measuring techniques, and the exposure and locations of instruments. Spurious changes may also arise from, for example, urban heat-island effects and coverage changes. Errors arising from these factors have been painstakingly minimised, but small residual uncertainties remain.

The most striking feature of this record is the overall warming trend, with the most recent years being the warmest. The record, however, shows a number of other important features. First, there are large variations from year to year. Some of these variations are associated with El Niño, a small number reflect short—term coolings due to volcanic eruptions, and the remainder are probably manifestations of the climate system's own internally generated variability. The record also shows large changes on the 10 to 30 year time scale. These probably reflect anthropogenic and solar forcing effects combined with internal variability.

Critics of the IPCC and the anthropogenic global warming hypothesis often point to the apparent discrepancy between the small greenhouse-gas forcing over 1910 -1940 and the rapid global warming that occurred during this period. It is true that this warming was too rapid to be accounted for by anthropogenic forcing alone. However, when the possible effects of internally generated variability and solar forcing are accounted for, there is no serious discrepancy.

Free Atmosphere Changes

Human influences on climate are not restricted to the surface. Simple physics demands that any anthropogenic warming should extend throughout the troposphere, primarily because the convective activity associated with clouds keeps this part of the atmosphere well mixed. Above the troposphere, both CO_2 and ozone-depletion effects should have led to cooling, especially in the lower stratosphere. In searching for evidence of human influences, there fore.

Temperatures above the Earth's surface have been measured since the 1940s. The longest records are those obtained from instruments carried aloft on weather balloons, which are reliable back to the early 1960s. For the troposphere, these data show an overall warming trend, the magnitude of which is very similar to the surface data trend. Over the same period, the data show a marked cooling in the stratosphere. Both the tropospheric warming and the stratospheric cooling are consistent with the predictions of climate models for the joint influences of increasing greenhouse-gas concentrations and halocarbon-induced stratospheric ozone depletion.

Since 1979, in addition to radiosonde data, a more spatially complete picture is available fro m space using Microwave Sounding Unit (MSU) instruments on weather satellites. Computer weather forecasting models have also been used in recent years to produce syntheses of data from different sources. In the troposphere, the different records show different trends. The satellite data show no significant trend, while some radiosonde data show a warming trend that is quite similar to the surface warming trend. In the stratosphere, all records a reconsistent in showing a marked cooling.

This difference in trends since 1979 between the satellite (MSU) data for the troposphere and the surface data has led some to proclaim that the surface data are flawed and, furthermore, that the lack of a significant MSU trend implies that model predictions of anthropogenic global warming are wrong. Both conclusions oversimplify what is, in fact, a very complex scientific issue. Tropospheric and surface data are different things, so one would not expect them to show identical trends over a period as short as 20 years.

The most obvious explanation for the difference is data uncertainties, which exist for both data sets. For surface data, as noted above, uncertainties arise through instrumentation changes, nonclimatic influences such as urban heat-island effects, and coverage changes and deficiencies. Careful quality control procedures have been applied to minimise these potential error sources.

Precipitation Changes

Precipitation is much more variable in both time and space than temperature, and reliable long-term records exist only over the Earth's land areas; and, even here, the coverage is incomplete. Changes in annual total precipitation averaged over the land are as of the globe from the Hulme data set. The dominant characteristic of this record is its marked year-to-year variability.

If smaller regions are examined, the year- to—year variability becomes even more pronounced. In the assessment of this record in the IPCC Second Assessment Report (SAR), it is stated (p. 156) that the precipitation data show a small positive trend, amounting to +1 percent per 100 years. Unfortunately, one cannot place much confidence in this early part of the record because of data quality problems and reduced spatial coverage. Thus, there is no firm evidence of any real overall trend.

Because of the high interannual variability in the precipitation record, associating regional—and/or global—scale precipitation changes with any specific causal mechanism is extremely difficult. Apart from changes in average precipitation levels, changes have also been observed in the distribution of precipitation amounts. An important example comes from North America. Here, Karl and colleagues have found that the frequency of extreme daily rainfall events has increased in recent times. They have also shown that the changes are more than one would expect to have occurred by chance. Further, they note that there are qualitative arguments to suggest that similar changes might occur because of greenhouse-gas-induced global warming. These are suggestive results, but they do not prove a cause effect relationship.

Designation and Detection

The IPCC Second Assessment Report states that "the balance of evidence suggests a discernible human influence on global climate". Why did the scientists who wrote the IPCC Second Assessment Report feel able to make such a statement, when, in the previous full IPCC report, they were unable to do so? The critical difference came through the availability of quantitative estimates of the climatic effects of anthropogenically produced sulphate aerosols. Both global-mean and regional-scale data have played important, but complementary roles in recent detection and attribution studies.

In 1990, it was noted that only the lowest estimates of anthropogenic warming based on model calculations were consistent with the observed changes in global-mean temperature. It was further noted that the pattern of observed temperature change did not match that expected to arise from increased greenhouse-gas concentrations based on general circulation model (GCM) results. The possibility that sulphate aerosols might account for these discrepancies was first raised in 1989. At the global-mean level, later calculations have shown that the inclusion of aerosol effects can improve the fit between models and observations. If both aerosol effects and the effects of solar forcing are considered, the model-predicted warming is in close agreement with the observations.

Including the effects of sulphate aerosols has also been shown also to improve the correspondence between model predictions and observed patterns of temperature change, both in the horizontal plane and in the vertical plane. These correspondences, based on rigorous statistical tests, are too close to have occurred by chance. Overall, therefore, there is good agreement between model predictions and observations at both the spatial-mean and spatial pattern levels.

These detection and attribution studies have employed only temperature data. The relative importance of human factors varies greatly according to both the spatial scale and the variable considered. As a general rule, the smaller the spatial scale, the smaller the ratio of humantonatural influences. Furthermore, the magnitude of the human influence relative to natural variability for temperature is, generally, much larger than for variables like precipitation and atmospheric circulation. These differences are important in understanding future changes.

Predicting Future Climate Changes

Future Emissions Scenarios

The starting point for predicting future changes in climate is usually a "scenario" defining future emissions and/or concentrations of a range of gases. If a scenario involves future emissions, then these must first be translated into future concentrations using appropriate models. The concentrations in turn determine how the balance between incoming short-wave and outgoing long-wave radiation will change; and changes in the radiation balance determine how the climate will change.

It is possible to distinguish two types of emissions scenarios: scenarios that do not explicitly include climate-related policies, and policy scenarios. The former, referred to here as "no-climate-policy" scenarios, give an idea of what might happen in the absence of new policies to limit climate change. Such scenarios are often referred to as "business-as-usual" (BAU) scenarios; but this can be a misleading term, not least because these no-climate-policy scenarios may include the effects of existing or projected policies to reduce other environmental problems such as air pollution and acid precipitation. This is particularly important for SO_2. Only no-climate-policy scenarios are considered here.

Future emissions of the gases that may affect climate depend on future changes in population, economic growth, energy efficiency, and evolving policies to limit emissions. Once these determinants ave been specified, they can be used in multi-disciplinary integrated assessment models to define future emissions scenarios. Because the determinants are uncertain, a wide range of emissions scenarios can be produced even in the absence of emissions limitations policies.

The IS92 scenarios have some well-recognised limitations. For this reason, and because a number of years have passed since they were constructed, a new set of no-climate-policy scenarios is being developed for an IPCC Special Report on Emissions Scenarios (SRES). Preliminary versions of four "marker" scenarios were released by the

SRES writing team in December 1998, for use by the international scientific community in climate model simulations that will, in turn, be used in the IPCC Third Assessment Report. These are referred to as the SRES A1, A2, B1, and B2 scenarios. It should be noted that, at the time of this writing, these scenarios have not yet been approved through the formal IPCC review process. They are, however, the most up-to-date and comprehensive emissions scenarios available. The four marker scenarios are used here with permission from the groups and individuals who produced them.

The most marked difference between the SRES scenarios and the earlier IS92 scenarios is in the emissions projections for SO_2. For this gas, the IS92 scenarios did not fully consider the effects of policies to combat air pollution and acid rain. The new SRES emissions scenarios include, in more realistic and internally consistent ways, the possible effects of such policies. In the IS92 scenarios, SO_2 emissions generally increase markedly—e.g., in IS92 a from 75 TgS/yr in 1990 to roughly double this in 2050. In contrast, the new SRES scenarios project eventual decreases in SO_2 emissions over the next century. Since SO_2 emissions lead to the production of sulphate aerosols, which have a strong cooling effect, climate projections based on the SRES scenarios are likely to differ markedly from those based on the IS92 scenarios.

Atmospheric Concentrations and Radiative Forcing

Given an emissions scenario, concentrations may be determined using models that relate changes in atmospheric concentration of a gas to the atmospheric inputs and outputs. Such models are referred to as gas-cycle models. The predicted concentrations may then be interpreted in terms of their radiative forcing consequences.

The values are similar to those for the IS92 scenarios, but, because the emissions range is smaller, they span a range in 2100 that is somewhat narrower than for the IS92 scenarios. These values are subject to uncertainties arising from uncertainties ability to model the carbon cycle. In terms of their climate consequences, however, these uncertainty effects are relatively small.

Total anthropogenic forcing is 1.07 W/m^2 over the 1765 -1990 period. For the future, forcing for the SRES scenarios from 1990 to 2050 ranges from 2.30 to 3.11 W/m^2, in all cases larger than IS92a. All SRES scenarios have CO_2 as the dominant forcing agent, all show important additional forcings due to the sum of other (non-CO_2) greenhouse gases, and all have a positive forcing contribution from sulphate aerosols from 1990 to 2100.

Global-Mean Climate Prominence

To project global-mean temperature changes, the model used is the upwelling—diffusion energy-balance model (UD EBM) of Wigley and Raper. The UD EBM also calculates the amount of expansion of the ocean water mass due to warming. The amount of warming—related melting from glaciers and small ice sheets and from Greenland and Antarctica is added to this to calculate changes in sea level. The approach used is the same as was used in the IPCC SAR. The only change is in using the preliminary SRES scenarios as the drivers for future change rather than the IS92 scenarios.

Coupled ocean/atmosphere general circulation models (O/AGCMs) nevertheless remain the "gold standard" for future climate simulations. Thus, a most important consideration in using simpler models such as UD EBMs is that they should accurately simulate the results of O/ AGCMs when used for the same experiments. This was the basis for the use of a UD EBM in the IPCC SAR.

Agreement between the simple model used by IPCC and O/AGCM results is demonstrated in Kattenberg et al. In essence, simple models are used as relatively sophisticated interpolation and extrapolation tools. They were used in the IPCC SAR to consider a wider range of scenarios than could practically be considered with O/AGCMs, and to assess the magnitude of uncertainties associated with, for example, uncertainties in the climate sensitivity.

Global-mean temperature and sea level results for the four SRES marker scenarios based on "best-estimate" model parameters. The full range of results spanning the scenarios and accounting for uncertainties in the climate sensitivity (DT2x) and, for sea level, uncertainties in the ice-melt model parameters. The global-mean warming fro m 1990 to 2100 ranges between 1.9°C and 2.9°C. Sea-level rise estimates over the same period for the four scenarios. The inter-scenario range is 46 to 58 cm. These temperature and sea level results are similar to the central estimates given in the IPCC SAR of 2.0°C and 49 cm.

From 1990 to 2100, the range of global-mean warming estimates is 1.3 -4.0°C. Global-mean sea-level rise over the same period is between 17 cm and 99 cm. The corresponding IPCC SAR ranges are 0.8—3.5°C and 13 - 94 cm. The values here are shifted up from those in the IPCC SAR because of the lower SO_2 emissions in the SRES scenarios. An important point to note is that the uncertainty range for the SRES scenarios is determined m o re by climate sensitivity and sea-level modelling uncertainties than by emissions uncertainties, especially for sea level.

Climate Change for the United States

The ideal tool to use for estimating the spatial details of future climate is the coupled ocean / atmosphere GCM

(O/AGCM). A number of simulations of future climate have been carried out with this type of model, but, to date, no work has been published in which the simulations use up-to-date combinations of future greenhouse gas and SO_2 emissions. Such simulations, based on the IPCC SRES scenarios, are currently being carried out by a number of GCM modelling groups for input into the Third Assessment Report.

Model Evaluation Procedure

How credible are currently available GCMs? There are two ways to answer this question. The first is a standard model evaluation procedure: one simply compares the model's simulation of current climate with observations. Analyses like these give widely varying results. Some models are good in one region and less good in another, and some models perform well for some variables but relatively poorly for others. A second approach is to compare the results of different models when they are all used to perform the same type of climate-change experiment.

For the present analysis, results from 15 different models are compared. The models considered are those compiled in the SCENGEN software package. These models have different vertical and horizontal resolutions and re p resent different model "vintages." Most of the models are MLO/AGCMs, but four are coupled O/AGCMs. The first part of the present model evaluation is to compare model simulations of present- day climate with observations.

Only a single criterion is used, the average of the global pattern correlation between modelled and observed precipitation. High values of this correlation indicate that modelled and observed precipitation patterns are similar, and low values point to important differences. A correlation of 0.707 is required for modelled and observed patterns to have 50 percent of their spatial variability in common. Only four models reach this threshold. If one plots this pattern correlation against model year, there is an upward trend pointing to improvements in the models over time.

The second part of the present model evaluation is to compare the results of different models for a similar climate change experiment. If all models are asked the same question, how well do the models agree? Lack of agreement would imply that there is considerable uncertainty regarding regional—scale climate-change results, and that one should be cautious in using results from any one model. Model agreement, of course, would not guarantee that their results were unequivocally correct.

The experiment used here for inter-model comparison is one where the CO_2 concentration is doubled. The data used were seasonal-mean changes in temperature and precipitation for winter ; spring ; summer; and fall. The comparisons show that some model pairs have very similar patterns of change, while other pairs give highly dissimilar results. The best results are obtained for winter temperature-change patterns, largely because many models show an enhanced warming in higher latitudes. The worst results are for summer and fall precipitation—change patterns. Here inter-model differences are generally very large.

For temperature, the modelled changes are always larger than any differences between the models. In other words, there is a clear warming signal over the whole region and in all seasons that is common to all models. For precipitation, the inter-model comparison results are less satisfactory. Generally, the average signal is smaller than the average difference between the models. This is particularly the case in summer and fall. There is a clearer signal in winter and spring in the northern 10° latitude band of the study area.

Future Climate Change Possibilities

The climate sensitivity is assumed to be DT2x = 2.5°C. Patterns of climate change for 2030, one simply reads the global-mean warming directly and scales the normalised patterns of change by 0.7. To obtain an absolute climate scenario, one would add these changes to the current climate. Sulphate aerosol effects will undoubtedly modify these results.

At the global-mean level, the forcing contribution from sulphate aerosols is small relative to the total forcing. However, because of the large spatial variability in the emissions of SO_2 and the forcing from sulphate aerosols, there may still be important effects at the regional level. These effects will vary with emissions scenario and time. At present, it is not possible to give any reliable indication of what they may be, partly because appropriate O/AGCM model experiments have yet to be performed also because of the very large uncertainties surrounding the quantification of the relationships between SO_2 emissions and the resulting forcing effects.

Other Aspects of Climate Changes

For the latter, only temperature and precipitation were considered. Over the United States, one can be fairly confident that the warming will be greater than the global-mean warming worldwide, with greatest enhancement at high latitudes in winter. For precipitation, the changes are far more uncertain, largely because different models give widely differing results. The only result common to most models is a precipitation increase in winter over the northern Great Plains/Great Lakes region, and northeastern states. In the central and southern latitude bands of the United States, some models show substantial increases in precipitation, while others predict substantial decreases. The impacts of climate change at any particular location will, however, be determined by factors other than just changes in mean temperature and precipitation.

Distribution of Temperatures

A general warming will shift the whole distribution of temperatures. Thus, relative to any fixed threshold, the frequency of warm temperature extremes will increase and the frequency of cold extremes will decrease. This is a general result, applicable to any part of the globe. In the absence of variability changes, the increase in the frequency of extreme warm events will be disproportionally large.

Variability Changes

Changes in variability are important because they may have a significant effect on agriculture and water resources. Furthermore, the IPCC Second Assessment Report notes that "a small change in variability has a stronger effect than a small change in the mean," as pointed out earlier by Wigley. There is, however, no consensus between models on changes in the interannual variability of climate elements like temperature and precipitation.

Indeed, even the best models perform poorly in simulating such variability—i.e., their simulations of current variability differ noticeably from observed variability. If any changes did occur, they would be regionally specific, so that some regions might experience increases in variability while nearby regions might experience changes in the other direction.

Precipitation Extremes Changes

The IPCC Second Assessment Report notes that GCM results suggest increases both in the frequency of intense precipitation events and, in some regions, in the probability of dry days and the length of dry spells. Two more recent studies support this conclusion. Zwiers and Kharin found that heavy precipitation events over North America might occur twice as often in a world that was 3.5°C warmer than today. Freietal. found a similar shift to more frequent heavy precipitation events in southern Europe. While these analyses are careful and comprehensive, one must still be cautious in accepting their quantitative conclusions.

In both studies, the warming considered is substantially greater than that expected over the next 50 years. As an additional cautionary note, Osborn has shown that one cannot automatically translate changes in precipitation intensity at the GCM gridbox level to real—world local changes. In some cases, in making this spatial-scale conversion, an increase in intensity can become a decrease.

Thus, while both types of change are possible in the United States, there is no unequivocal evidence for either. Since warming should lead to increased evaporation, if precipitation were not to change at all at a particular location, soil moisture levels and the availability of water for runoff would have to decrease. However, even this conclusion is subject to uncertainty because of the direct plant-physiological effect of increasing CO_2 concentrations on plant water-use efficiency.

If, as small-scale experiments suggest, water- use efficiency increases with increasing CO_2, then plants would transpire less in the future. To some degree, at least, this would offset any tendency toward increased evaporation as a result of warming. The big uncertainties here are in scaling up the small-scale experimental results to larger, ecosystem scales, and in knowing how ecosystems will respond to future time-varying changes in climate.

Midlatitude Storms Systems

For midlatitude storm systems, the state of science is exemplified by IPCC's cautious statement that "... there is little agreement between models on... changes in storminess... (and) conclusions regarding extreme events are obviously even more uncertain".

Tropical Storms

The formation of tropical storms is controlled by many different factors, including sea surface temperatures, atmospheric stability, wind shear, the large scale circulation in which a storm may be embedded, and high-level wind patterns. Current GCMs used in climate studies do not have fine enough spatial resolution to be able to simulate individual tropical cyclones. Furthermore, even the most sophisticated weather forecasting models are generally unable to predict the initiation of tropical cyclones.

Nevertheless, there is empirical evidence that there might be small increases in the frequency of Atlantic hurricanes, based on the positive correlation between SSTs and hurricane frequencies in this region. There is also model evidence that minimum pressures may decrease and wind speeds may increase in tropical storms worldwide.

Knutson et al., for example, project wind speed increases of 5 to 12 percent for a sea-surface temperature increase of 2.2°C. However, the projected changes are small relative to past interannual variability. Thus, even if these projections could be considered reliable, it would be many decades before the hypothesized signals could be positively detected above the noise of interannual variability.

An associated possibility is that, along with a minor intensity increase, there could be substantially larger changes in the amount of precipitation associated with individual storms. This may be a more robust result because, with increased ocean temperatures, it is almost certain that the moisture-holding capacity of the atmosphere will increase. Along with this, one would expect increased precipitation at the global-mean level. While the manifestation of this general increase over midlatitude land areas is highly uncertain, more confidence can be placed on the possibility of precipitation increases in areas currently frequented by tropical cyclones.

2

Causes of Global Climate Change

The global climate must be viewed as operating within a complex atmosphere/earth/ocean/ice/land system. Any change to this system, resulting in climate change, is produced by forcing agents—the causes of climate change. Such forcing agents may be either internal or external. External forcing mechanisms involve agents acting from outside the climate system. By contrast, internal mechanisms operate within the climate system itself.

Any change in the climate must involve some form of energy redistribution within the global climate system. Nevertheless, forcing agents which do not directly affect the energy budget of the atmosphere (the balance between incoming solar radiation and outgoing terrestrial radiation, are considered to be non-radiative mechanisms of global climate change. Such agents usually operate over vast time scales and mainly include those which affect the climate through their influence over the geometry of the Earth's surface, such as location and size of mountain ranges and position of the ocean basins.

Radiative forcing Mechanism

A process which alters the energy balance of the Earth-atmosphere system is known as a radiative forcing mechanism. These may include variations in the Earth's orbit around the Sun, solar radiation, volcanic activity and atmospheric composition. Associating a particular cause with a particular change, however, is extremely difficult. The interlinked nature of the climate system ensures that there are feedbacks; a change in one component leads to a change in most, if not all, other components. Before investigating some of the more important forcing mechanisms, both internal and external, there is one factor that needs elaborating: time scale.

Time Scales investigating Climate Variation

The importance of considering different time scales when investigating climate change has already been identified. Climate varies on all time scales, in response to random and periodic forcing factors. Across all time periods from a few years to hundreds of millions of years there is a white noise of random variations of the climate, caused by internal processes and associated feedback mechanisms, often referred to as stochastic or random mechanisms. Such randomness accounts for much of the climate variation, and owes its existence to the complex and chaotic behaviour of the climate system in responding to forcing. An essential corollary of the existence of random processes is that a large proportion of climate variation cannot be predicted.

Of far more relevance are the periodic forcing factors, for by understanding their mechanisms and the impacts they have on the global climate, it is possible to predict future climate change. How the climate system responds to periodic forcing factors, however, is often not clear. If it is assumed that the climate system responds in a linear fashion to periodic forcing, variations in climate will exhibit similar periodicity. If, however, the response of the system to forcing is strongly non-linear, the periodicities in the response will not necessarily be identical to the periodicities in the forcing factors. Frequently, the climate responds in a fashion intermediate between the two.

There are many climate forcing factors spanning an enormous range of periodicities. The longest, 200 to 500 million years, involves the passage of Solar System through the galaxy, and the variations in galactic dust. These may be considered to be external forcing mechanisms. Other long time scale variations include the non-radiative forcing mechanisms, such as continental drift, orogeny and isostasy. These are internal forcing mechanisms. External changes in the amount of solar radiation and the Earth's orbit around the Sun, and internal variations in volcanic activity, ocean circulation and atmospheric composition, all occur over time scales from 1 year to 10^5 years. Additionally, there are numerous other internal feedback mechanisms which all contribute to the changing of the global climate.

The actual climate state at any point in time represents an aggregate response to all cycles of variation superimposed on the background noise. The response of the climate system to this combination of forcing factors itself depends upon the different response times of the various components of the system. The overall climatic response will then be determined by the interactions between the components. The atmosphere, surface snow and ice, and surface vegetation typically respond to climatic forcing over a period of hours to days. The surface ocean has a response time measured in years, whilst the deep ocean and mountain glaciers vary only over a period spanning hundreds of years. Large ice sheets advance and withdraw over thousands of years whilst parts of the geosphere respond only to forcing periods lasting hundreds of thousands to millions of years.

The response of the climate system to episodes of forcing can be viewed as a form of resonance. When the time period of forcing matches most closely the response time of a particular system component, the climatic response will be greatest within that component. Milankovitch forcing, with periods of tens of thousands of years will be manifest in the response of the ice sheets, and the overall response of the climate system will be dominated by changes within the cryosphere.

In addition, longer response times of certain components of the climate system modulate, through feedback processes, the short term responses. Throughout the remainder of this matter, it should be recognised that a range of time scales applies to climate forcing mechanisms, radiative and nonradiative, external and internal, and to the response of the different components of the climate system. The various external forcing mechanisms operating over time scales of 10 years to 10^9 years.

Effect of Galactic Variations

The orbit of the Solar System about the centre of the Galaxy has been considered as a possible external climate forcing mechanism. During the course of a galactic year, variations in the interstellar medium may influence the amount of solar radiation incident at the Earth's surface, thus acting as a radiative forcing mechanism to induce climate change. Some argues that variations in gravitational torque induced by Galaxy's near neighbours, the Small and La-rge Magellanic Clouds, could have far-reaching consequences for the Earth's cli-mate.

Unfortunately, the enormous time scale associated with this forcing (and any hypothesised global climatic change) makes empirical confirmation of this premise exceedingly imprecise. Nevertheless, it is indeed possible that the ice age supercycles during the last 700 million years could be the result of such galactic forcing mechanisms.

Effect of Orbital variations

In the mid-19th century, Croll proposed an astronomical theory linking the Pleistocene (2 Ma to 10 Ka) ice ages with periodic changes in the Earth's orbit around the Sun. Croll's ideas were later refined and elaborated by Milankovitch. The Milankovitch theory is the name given to the astronomical theory of climate variations. Since these ideas were put forward, much evidence has been found to support the theory.

The original Milankovitch theory identifies three types of orbital variation which could act as climate forcing mechanisms, obliquity or tilt of the Earth's axis, precession of the equinoxes and eccentricity of the Earth orbit around the Sun. Each variation has its specific time period.

Obliquity: Today the Earth is tilted on its rotational axis at an angle of 23.4° relative to a perpendicular to the orbital plane of the Earth. Over a 41,000 year time period, this angle of inclination fluctuates between 22° and 24.5°, influencing the latitudinal distribution of solar radiation. Obliquity does not influence the total amount of solar radiation received by the Earth, but affects the distribution of insolation in space and time.

As obliquity increases, so does the amount of solar radiation received at high latitudes in summer, whilst insolation decreases in winter. Changes in obliquity have little effect at low latitudes, since the strength of the effect decrease towards the equator. Consequently, variations in the Earth's axial tilt affect the strength of the latitudinal temperature gradient. Increased tilt has the effect of raising the annual receipt of solar energy at high latitudes, with a consequent reduction in the latitudinal temperature gradient.

Eccentricity: The Earth's orbit around the Sun is not perfectly circular but follows an elliptical path. A second orbital variation involves the strength of the ellipse, or eccentricity. This parameter, *e*, which compares the two focal lengths, *x* and *y* in

$$e = \{ (x^2 - y^2)^{1/2} \} / x$$

When the orbit is circular, the lengths *x* and *y* are equal and $e = 0$. The Earth's orbit has been found to vary from being near circular ($e = 0.005$) to markedly elliptical ($e = 0.06$) with two primary periodicities of approximately 96,000 and 413,000 years. The current value of e is 0.018. Variations in eccentricity influence the total amount of solar radiation incident at the top of the Earth's atmosphere.

Precession has two components: an axial precession, in which the torque of the other planets exerted on the Earth's equatorial bulge causes the rotational axis to gyrate like a spinning top; an elliptical precession, in which the elliptical orbit of the Earth itself rotates about one focus. The net effect describes the precession of the equinoxes with a period of 22,000 years. This term is modulated by eccentricity which splits the precession into periods, 19,000 and 23,000 years. Like obliquity, precession does not affect the total amount of solar energy received by the Earth, but only its hemispheric distribution over time. If the perihelion occurs in mid-June i.e. when the Northern Hemisphere is tilted toward the Sun, then the receipt of summer solar radiation in Northern Hemisphere will increase. Conversely, if the perihelion occurs in December, the Northern Hemisphere will receive more solar radiation in winter. It should be clear that the direction of changes in solar radiation receipt at the Earth's surface is opposite in each hemisphere.

Milankovitch cycles and ice ages: The three components of the orbital variations together effect both the total flux of incoming solar radiation and also the temporal and spatial distribution of that energy. These variations have the potential to influence the energy budget of the climate system, and can therefore be regarded as possible causes of climate change over a 10^4 to 10^5 year time scale. Being external to the climate system, they may be classified as external forcing mechanisms.

Milankovitch considered the changing seasonal (precession) and latitudinal (obliquity) patterns of incoming radiation to be critical factors in the growth of continental ice sheets and in the initiation of ice ages. He hypothesised that when axial tilt was small, eccentricity was large and perihelion occurred during the Northern Hemisphere winter, such a configuration would allow the persistence of accumulated snow throughout the summer months in the Northern Hemisphere. Additionally, the warmer winters and stronger atmospheric general circulation due to the increased temperature gradient would increase the amount of water vapour at the high latitudes available for snowfall. For long-term proxy temperature data, spectral analysis, which permits the identification of cycles, has shown the existence of periodicities of 100,000, 43,000, 24,000 and 19,000 year, all of which correspond closely with the theoretical Milankovitch cycles.

Nevertheless, verification of a causal link between the orbital forcing factors and the climatic response is far from being proved, and significant problems remain.

This would be the result of eccentricity variations in the Earth's orbit, which alone account for the smallest insolation changes. Secondly, it is not clear why changes in climate appear to be global. A priori reasoning indicates that the effects of precession would cause opposite responses in each hemisphere. In fact, climate change is synchronised between Southern and Northern Hemispheres, with a growth of ice sheets during glaciations occurring in the Arctic and Antarctic. It is now widely believed that the circulation of the oceans provides the forcing factor for synchronisation.

Most crucially of all, however, it seems that the orbital forcing mechanisms alone, could not account for the observed climatic variations over the past 2 million years. In order to explain the magnitude of the observed climatic changes, it seems necessary to invoke various feedback mechanisms. Indeed, Milankovitch himself had expected the direct effects of variations in insolation to be magnified by feedback processes, such as, at high latitudes, the ice albe do effect.

Effect of Solar Variations

Although solar variability has been considered, a priori, to be another external forcing factor, it remains a controversial mechanism of climate change, across all time scales. Despite many attempts to show statistical associations between various solar periodicities and global climate cycles, no realistic causal mechanism has been proposed to link the two phenomena. The best known solar cycle is the variation in the number of sunspots over an 11 year period. Sunspot cycles are thought to be related to solar magnetic variations, and a double magnetic cycle (approximately 22 years) can also be identified. What is of interest to a climatologist is whether the sunspot cycles are accompanied by variations in solar irradiance—the solar constant—which, potentially, could force climate changes. The solar constant (approximately 1368 Wm^{-2}) is a measure of the total solar energy flux integrated across all wavelengths of radiation. Two decades of satellite observations reveal that the solar constant varies over time scales of days to a decade, and there does appear to be a significant relationship with the sunspot number cycle. At times of high sunspot number, the value of the solar constant increases.

Although sunspots are regions of cooler than average Sun surface temperature, their presence is accompanied by brighter (hotter) faculae which more than compensates for the increase in darker sunspot areas. This relationship can be extended back over time using the long sunspot record. The difficulty in attributing any observed climate change to these variations in solar irradiance is that the latter are small in magnitude—a change of much less than 1% over the course of the sunspot cycle. Wigley stressed that with such small variations in the solar constant, the global climatic response would be no more than a 0.03o C temperature change.

Nevertheless, many climatic records (e.g. indices of droughts, temperature and total atmospheric ozone) do appear, at least statistically, to display periodicity linked to one or both of the sunspot cycles. It should be clear, however, that a statistical association between solar variability and climate change does not prove cause and effect.

It is of course possible that the approximate 11-year cycle identified in many climate records is caused by some unknown internal oscillation and not by external solar forcing. It is conceivable that, simply by chance, the phase of the oscillation could coincide with the phase of the solar variability. More plausibly, an internal oscillation can become phase-locked to the solar cycles, thus augmenting the climatic response by a kind of feedback mechanism. For the time being, therefore, the link between the sunspot cycles and climate change must remain a speculative one.

However, there are other solar periodicities, with longer time scales that could be considered as climate forcing mechanisms. It has been suggested that the long-term variation in the amplitude of the sunspot cycles may have an influence on global climate. As with the sunspot cycles, however, the evidence is largely circumstantial. Other solar variations include cycles of sunspot cycle length, changing solar diameter and the rate of change of solar diameter. Although some of these long-term variations may involve larger changes in solar output, this is again mere speculation.

Proxy records of solar irradiance changes are needed when even longer time scales are considered. A number of scientists have used records of ^{14}C in tree rings to investigate the relationships between potential solar forcing mechanisms and climate change. Changes in the output of energetic particles from the Sun are believed to modulate the production of ^{14}C in the upper atmosphere. The magnetic properties of the solar wind change with the variation of sunspots, leading in turn to variations in the production of ^{14}C. The effect of the solar wind is such that high ^{14}C production is associated with periods of low sunspot number.

Relatively long and reliable ^{14}C records are now available. Spectral analysis has revealed a number of solar periodicities including a 2,400 year cycle, a 200 year cycle, a 80 to 90 year cycle and the shorter 11 and 22 year cycles. The ^{14}C records have also been correlated with a number a climate change indicators, including glacial advance-retreat fluctuations and annual temperatures for England. Episodes of low ^{14}C production are associated with high sunspot activity and warmer climates. It is certainly feasible that the climatic variations of the Holocene, and the shorter fluctuations associated with the Little Ice Age have been forced by the interacting millennia and century scale cycles of solar activity. However, conclusive evidence of a mechanism linking cause and effect is again missing.

Use of Internal forcing mechanisms

Some of the various internal forcing mechanisms operating over time scales of 1 year to 10^8 years. They may be either radiative or non-radiative forcing mechanisms.

Orogeny is the name given to the tectonic process of mountain building and continental uplift. Such mechanisms operate only over tens or even hundreds of millions of years. The Earth's outer surface, a layer known as the lithosphere, is broken up into about 12 different plates which are constantly adjusting their positions relative to each other. Such movements are driven by the internal convective dynamics within the Earth's mantle. When plates collide, one may either be subducted beneath another, or both are pushed continually together, forcing upwards any continental land masses, to form long mountain ranges. The Himalayas formed when the Indian plate crashed into Asia about 20 to 30 million years ago.

There is now little doubt that the presence of mountain ranges on the Earth can dramatically influence global climate, and that orogenic uplift can act as a non-radiative forcing mechanism. North-south orientated mountain ranges in particular have the ability to influence global atmospheric circulation patterns, which usually maintain a more east-west trend on account of the Coriolis Force.

During chemical weathering, carbon dioxide is extracted from the atmosphere to react with the decomposing rock minerals to form bicarbonates. These bicarbonates are soluble and can be transported via rivers and other fluvial channels, finally to be deposited on ocean floors as sediment. In essence, carbon dioxide is sequestered from the atmosphere, thereby decreasing the Earth's natural greenhouse effect, causing further cooling. In view of this greenhouse feedback, mountain uplift seems to generate both non-radiative forcing and radiative forcing. In such situations as described above, isolating a primary cause of climatic change from its secondary feedbacks, becomes ineffective. Mountain uplift may also increase the land surface area covered by snow the year round. The subsequent increase in planetary albedo will reduce the amount of energy absorbed at the Earth's surface, initiating further cooling.

Epeirogeny is the term used to describe changes in the global disposition of land masses, and like orogenic processes, these changes are driven by internal plate tectonic movements. Because the internal dynamics of the Earth are slow, continents move about the globe at a rate of several centimetres per year. However, over tens or hundreds of millions of years, both the size and position of land area can change appreciably. At times in Earth history, there have been super continents in which all the continental plates were locked together in one area of the globe. The last of these occurred about 250 million years ago, and is named Pangea. Since that time, the continents have gradually moved apart, the most recent separation occurring between Europe and North America, during the last 60 to 70 million years. What is now the Pacific Ocean used once to be the vast expanse of water, called the Panthalassa Ocean, that surrounded Pangea. A number of possible mechanisms which forced global climate to fluctuate between "greenhouse" and "icehouse" states have been explored.

As the continental area occupying high latitudes increases, as a result of continental drift, so the land area with permanent ice cover may expand, thus raising the planetary albedo, forcing (radiatively) a global cooling.

The arrangement of continental land masses significantly affects the surface ocean circulation. Since ocean circulation is involved in the latitudinal heat transport regulating global climate, so the wandering of land masses may force (non-radiatively) climate change over times scales involving tens or hundreds of millions of years.

Such long term variations in ocean circulation as a result of continental drift, in addition to orogenic processes, may have accounted for the return to a global "icehouse" that has taken place over the last 40 million years. Associated with continental drift is the tectonic process of sea floor spreading. It was explained how tectonic plates collide with one another and are consumed either by subduction or mountain building. New lithospheric plate material is formed at mid-ocean ridges, tectonic spreading centres, that mark the boundary between two diverging plates. These sea-floor regions, for example the Mid-Atlantic Ridge, release large amounts of energy and associated greenhouse gases.

The resulting ocean bathymetry is shallower that it otherwise would be and causes an rise in sea level. During Cretaceous times, midocean ridges were indeed more active than they are today. Consequently, sea levels stood several hundred metres higher (due also to the absence of waterstoring ice sheets), covering vast continental areas with shallow-level (epeiric) seas.

Such a situation may have two important consequences. First, ocean circulation will be markedly affected, influencing global climate as illustrated above. Second, the large shallow seas, with relatively lower albedos than the land areas which they submerge, would be capable of storing considerably more energy, thus heating the Earth's surface.

Explosive eruptions can inject large quantities of dust and gaseous material into the upper atmosphere, where sulphur dioxide is rapidly converted into sulphuric acid aerosols. Whereas volcanic pollution of the lower atmosphere is removed within days by the effects of rainfall and gravity, stratospheric pollution may remain there for several years, gradually spreading to cover much of the globe.

The volcanic pollution results in a substantial reduction in the direct solar beam, largely through scattering by the highly reflective sulphuric acid aerosols. This can amount to tens of percent. The reduction, is however, compensated for by an increase in diffuse radiation and by the absorption of outgoing terrestrial radiation. Overall, there is a net reduction of 5 to 10% in energy received at the Earth's surface. Clearly, this volcanic pollution affects the energy balance of the atmosphere whilst the dust and aerosols remain in the stratosphere.

Observational and modelling studies of the likely effect of recent volcanic eruptions suggest that an individual eruption may cause a global cooling of up to 0.3°C, with the effects lasting 1 to 2 years. Such a cooling event has been observed in the global temperature record in the aftermath of the eruption of Mount Pinatubo in June 1991.

The climate forcing associated with individual eruptions is, however, relatively short-lived compared to the time needed to influence the heat storage of the oceans. The temperature anomaly due to a single volcanic event is thus unlikely to persist or lead, through feedback effects, to significant long-term climatic changes. Major eruptions have been relatively infrequent this century, so the long-term influence has been slight. The possibility that large eruptions might, during historical and prehistorical times, have occurred with greater frequency, generating long-term cooling, cannot, however, be dismissed. In order to investigate this possibility, long, complete and well-dated records of past volcanic activity are needed.

One of the earliest and most comprehensive series is the Dust Veil Index (DVI) of Lamb, which includes eruptions from 1500 to 1900. When combined with series of acidity measurements in ice cores (due to the presence of sulphuric acid aerosols), they can provide valuable indicators of past eruptions. Using these indicators, a statistical association between volcanic activity and global temperatures during the past millennia has been found. Episodes of relatively high volcanic activity (1250 to 1500 and 1550 to 1700) occur within the period known as the Little Ice Age, whilst the Medieval Warm Period (1100 to 1250) can be linked with a period of lower activity.

Some argues that a link between longer time scale volcanic variations and the climate fluctuations of the Holocene (last 10,000 years). However, whilst empirical information about temperature changes and volcanic eruptions remains limited, this, and other suggested associations discussed above, must again remain speculative. Volcanic activity has the ability to affect global climate on still longer time scales. Over periods of millions or even tens of millions of years, increased volcanic activity can emit enormous volumes of greenhouse gases, with the potential of substantial global warming. However, the global cooling effects of sulphur dioxide emissions will act to counter the greenhouse warming, and the resultant climate changes remain uncertain. Much will depend upon the nature of volcanic activity. Basaltic outpourings release far less sulphur dioxide and ash, proportionally, than do the more explosive (silicic) eruptions.

The oceans store an immense amount of heat energy, and consequently play a crucial role in the regulation of the global climate system. In order to explain the observed hemispheric synchroneity of glaciation, despite periods of directly opposed orbital forcing in the two hemispheres, many researchers have looked to the oceans. Although, in this sense, changes in ocean circulation could be regarded as feedback resulting from orbital forcing, ocean circulation has traditionally been viewed as an internal forcing mechanism in its own right. At present, northern maritime Europe is warmed by heat carried polewards by the Gulf Stream. When the warm water meets cold polar air in the North Atlantic, heat is released to the atmosphere and the water cools and sinks. This is assisted by the increases in salinity that occur when sea ice forms in the Arctic regions. The bottom water so formed, called the North Atlantic Deep Water (NADW), flows southward through the western Atlantic, round Southern Africa and Australia, and then northwards into the Pacific Ocean. The North Atlantic is warmer than the North Pacific. The increased evaporation there therefore serves to increase salinity, relative to the North Pacific.

A number of the theories which have been put forward concerning the role of the oceans in the processes of climate change invoke changes in the rate of NADW production and other characteristics of the thermohaline circulation. Most attention has focused on the climatic transitions between glacial and interglacial episodes. It has been suggested that during a glacial period, the formation of the NADW is much reduced or even totally shut down. At these times, the Arctic ice sheets extends much further south into the North Atlantic, pushing the position of the polar front southwards. Cooler sea surface temperatures reduce evaporation and therefore salinity, further precluding the initiation of a thermohaline circulation. The concomitant absence of the warm surface Gulf Stream could result in northern Europe being 6 to 8°C colder than during interglacial times. The causes of the changes between the glacial and interglacial patterns of thermohaline circulation would then be seen as internal climate forcing mechanisms.

Indeed, Broecker has proposed that salinity changes between the North Atlantic and North Pacific may be so great that the entire global thermohaline circulation could be reversed. Such a theory of mode changes was developed in order to explain the rapid (<1,000 years) post-glacial climatic fluctuation of the Younger Dryas event about 11,000 years ago, when the North Atlantic appeared to cool by several degrees. Modelling appears to confirm the existence of at least two stable states of the thermohaline circulation. Rapid transitions between these two states, and the corresponding climatic flips between glacial and interglacial periods, in response to internal forcing, would then be nonlinear. Nevertheless, empirical evidence in support of mode changes is still inconclusive. Broecker concedes, however, that the shutdown of the North Atlantic 'heat conveyor belt system' alone would not be sufficient to initiate global temperature changes and ice sheet development.

Other internal feedback mechanisms would need to be invoked, for example changes in the concentration of greenhouse gases and aerosol loading, together with reduced ocean heat transport and increased ocean alkalinity. From the foregoing discussion on ocean circulation, one could conclude that such a mechanism of climate change should really be regarded as non-radiative, since what is at issue here is the transfer of energy within the ocean component of the climate system only. It is perhaps the resultant feedback processes identified in the preceding paragraph that allow most scientists to regard this mechanism as a radiative one.

The changing composition of the atmosphere, including its greenhouse gas and aerosol content, is a major internal forcing mechanism of climate change. The

Earth's natural greenhouse effect plays an important role in the regulation of the global climate. Obviously, then, changes in the atmospheric concentrations of greenhouse gases will modify the natural greenhouse effect, and consequently affect global climate. Changes in the greenhouse gas content of the atmosphere can occur as a result of both natural and anthropogenic factors, the latter which has received considerable attention in the last 20 years. Mankind, through the burning of fossil fuels, forest clearing and other industrial processes, has increased the amount of carbon dioxide and other greenhouse gases since the eighteenth century. Natural changes in greenhouse gas concentrations can occur in numerous ways, most often in response to other primary forcing factors. In this sense, as with ocean circulation changes, such forcing should be more strictly regarded as secondary forcing or feedback.

Changes in atmospheric CO_2 and methane (CH_4) have been associated with transitions between glacial and interglacial episodes. Much of the empirical evidence suggests that these changes lag behind the climate signal, and must therefore act as feedback mechanisms to enhance climate change rather than as primary forcing mechanisms. Changes in the atmospheric content of aerosols, again both natural and anthropogenic can act as climate forcing mechanisms, or more usually secondary feedback mechanisms. Increases in atmospheric turbidity will affect the atmospheric energy budget by increasing the scattering of incoming solar radiation. Atmospheric turbidity has been shown to be higher during glacial episodes than in interglacials, with a consequent reduction in direct radiation reaching the Earth's surface.

3

Ozone Depletion

Intriguingly, atmospheric ozone is not part of the planet's original system but a product of life on Earth, which began around 3.5 billion years ago. Until a half billion years ago, living organisms could not inhabit the land surface. Life was confined to the world's oceans and waterways, relatively protected from the intense unfiltered solar ultraviolet radiation. About 2 billion years ago as photosynthesising organisms emitted oxygen (O_2), a waste gas (ozone-O_3) gradually began to form within the atmosphere. From around 400 million years ago aqueous plants were able to migrate onto the now-protected land and evolve into terrestrial plants, followed by animal life that ate the plants. So the succession has evolved, via several evolutionary paths, through herbivorous and carnivorous dinosaurs, mammals and omnivorous humans. Today, terrestrial species are shielded by Earth's recently acquired mantle of ozone in the stratosphere that absorbs much of the solar ultraviolet.

Montreal Protocol-noticing and Responding to Ozone Depletion

Ozone depletion is one of several factors, including cloud cover and solar elevation, which affect ground level UV radiation. An examination of atmospheric changes in Australia from 1979 to 1992 has shown that the deseasonalised time series of UVR exposures were a linear function of ozone and cloud cover anomalies. In tropical Australia a trend analysis indicated a significant increase in UVR, estimated from sateilite observations, of 10% per decade in summer associated with reduced ozone and reduced cloud cover. In southern regions, a significant trend for UVR over time was not observed, partially due to increased cloud cover. Thus, in Tasmania, despite a significant ozone reduction of 2.1% per decade, measures of ground level UVR have not increased.

Estimating the resultant changes in actual ground-level ultraviolet radiation remains technically complex. Further, the methods and equipment used mostly have not been standardised either over place or time. While there is good agreement between similarly calibrated spectroradiometers, this may not be true when comparing different types of instruments-spectroradiometers, broad-band meters, filter radiometers. There is little or no reliable evidence on levels of UV radiation prior to concerns related to ozone depletion due to maintenance and calibration difficulties with these older instruments. The advent of satellite measuring systems allowed reliable measurement of UVR. However, satellite measurements may not accurately reflect ground level UVR due to failure to take adequate account of lower atmospheric changes.

It is clear that under cloud-free skies there is a strong correlation between ground level erythemal UV radiation and levels of atmospheric ozone. Yet the effects of clouds, increasing tropospheric ozone and aerosol pollution of the lower atmosphere modify this relationship making the detection of long-term trends in UVR related to ozone depletion difficult to elucidate. Long-term predictions are uncertain since they involve assumptions about not only future ozone levels but also future variations in cloud cover, tropospheric ozone and lower atmospheric pollution.

Fears of ozone depletion due to human activities first emerged in the late 1960s. A decade of denial and debate followed with eventual acceptance by scientists and policy-makers that ozone depletion was likely to occur and would represent a global environmental crisis. In the mid-1980s governments responded with alacrity to the emerging problem of ozone destruction. The Montreal Protocol of 1987 was adopted, widely ratified and the phasing out of major ozone-destroying gases began. The protocol was tightened further in the 1990s. At first sight, the solution to this particular global environmental change appears to be unusually simple: a substitution of particular industrial and agricultural gases for others.

Stratospheric Ozone Depletion and human-enhanced Greenhouse Effect

Stratospheric ozone destruction is an essentially separate process from greenhouse gas (GHG) accumulation in the lower atmosphere, although there are three important and interesting connections.

1. several of the anthropogenic greenhouse gases are also ozonedepleting gases.
2. tropospheric warming apparently induces stratospheric cooling that exacerbates ozone destruction. As more of Earth's radiant heat is trapped in the lower atmosphere, the stratosphere cools further, enhancing the catalytic destruction of ozone.
3. depletion of stratospheric ozone and global warming due to the buildup of greenhouse gases interact to alter UVR related effects on health. In a warmer world,

patterns of personal exposure to solar radiation are likely to change, resulting in increased UVR exposure.

This may be offset by changes in cloud cover and cloud optical thickness as a result of global climate change. Predictions of future UVR exposures based on ozone depletion, behavioural changes and climate change are uncertain. A recent analysis of trends in Europe reports a likely increase of 5-10% in yearly UV doses received over the past two decades.

Stratospheric ozone depletion has further indirect health effects. One important effect is that ozone depletion in the stratosphere increases the formation of photochemical smog, including ozone accumulation, in the lower troposphere. That is, ozone depletion in the upper atmosphere will allow more ultraviolet radiation to reach the troposphere where photochemical smog forms via a UVR mediated breakdown of nitrogen dioxide and other products.

Photochemical smog is a complex chemical mixture containing nitric acid (HNO_3); peroxyacyl nitrates (PANs), aldehydes ozone (O_3) and other substances. It has been estimated that the concentration of tropospheric ozone has increased from 10 ppb 100 years ago to 20-30ppb in some locations today, with peaks of >100ppb reported in some centres. The ozone component of photochemical smog acts as a respiratory irritant, causing oxidant damage to the respiratory epithelium and possibly enhancing allergen-induced airway inflammation.

Measurement of Solar UVR

Sunlight consists of solar rays of differing wavelengths. Visible light ranges from 400nm (violet) to 700nm (red). Infrared radiation, or heat, has longer wavelengths than visible light; ultraviolet radiation has shorter wavelengths than visible light. UVR is further divided into UVA (315-400nm), UVB (280-315nm) and UV-C (<280nm). Almost all incoming solar UVC and 90% of UVB are absorbed by stratospheric ozone, while most UVA passes through the atmosphere unchanged.

Although UVA penetrates human skin more deeply than UVB, the action spectra from biological responses indicate that it is radiation in the UVB range that is absorbed by DNA-subsequent damage to DNA appears to be a key factor in the initiation of the carcinogenic process in skin.

The amount of ambient UVB experienced by an individual outdoors with skin exposed directly to the sky is dependent on the following:

— stratospheric ozone levels

— solar elevation

— regional pollution

— altitude of the individual

— cloud cover

— presence of reflective environmental surfaces such as water, sand or snow.

The amount of received UVR exposure can be measured in terms of the energy of the transmitted photons, often expressed as energy per unit area irradiated.

To examine the health effects of solar UVR, it is necessary also to consider measurement in the biological dimension. Hence, UVR also is described in units of erythemal efficacy. To this end, exposure is spectrally weighted over the relevant wavelengths according to erythemal impact. Thus, standard erythemal doses (SEDs) can be defined by which daily, monthly or annual UV exposures can be quantified. A UV index also has been defined to express the daily maximum in biologically effective UVR, reached around midday.

Health Impacts of Stratospheric Ozone Depletion

There is a range of certain or possible health impacts of stratospheric ozone depletion. Many epidemiological studies have implicated solar radiation as a cause of skin cancer in fair-skinned humans. The most recent assessment by the United Nations Environment Programme projected significant increases in skin cancer incidence due to stratospheric ozone depletion. The assessment anticipates that for at least the first half of the twenty-first century additional ultraviolet radiation exposure will augment the severity of sunburn and incidence of skin cancer.

High intensity UVR also damages the eye's outer tissues causing "snow blindness", the ocular equivalent of sunburn. Chronic exposure to UVR is linked to conditions such as pterygium. UVB's role in cataract formation is complex but some subtypes, especially cortical and subcapsular cataracts, appear to be associated with UVR exposure while others do not. In humans and experimental animals, UVR exposure causes both local and whole-body immunosuppression. Cellular immunity is affected by variation in the ambient dose of UVR. UVR-induced immunosuppression therefore could influence patterns of infectious disease and may also influence the occurrence and progression of various autoimmune diseases.

Skin Damage

Since the 1850s it has been known that excessive exposure to sunlight can cause skin damage. Observation of boatmen, fishermen, lightermen, agricultural labourers and farmers revealed that skin cancer developed on areas most frequently exposed. The exact process by which exposure to sunlight causes skin cancer was not understood until relatively recently.

The incidence of skin cancer, especially cutaneous malignant melanoma, has been increasing steadily in white populations over the past few decades. This is particularly evident in areas of high UVR exposure such as South Africa, Australia and New Zealand. Human skin pigmentation has evolved over hundreds of thousands of years, probably to meet the competing demands of protection from the deleterious effects of UVR and maximisation of the beneficial effects of UVR. Skin pigmentation shows a clear, though imperfect, latitudinal gradient in indigenous populations. Over the last few hundred years, however, there has been rapid migration of predominantly European populations away from their traditional habitats into areas where there is a mismatch of pigmentation and UVR.

The groups most vulnerable to skin cancer are white Caucasians, especially those of Celtic descent living in areas of high UVR. Further, behavioural changes particularly in fair-skinned populations, have led to much higher UV exposure through sunbathing and skin-tanning. The marked increase in skin cancers in these populations over recent decades reflects, predominantly, the combination of post-migration geographical vulnerability and modern behavioural patterns. It remains too early to identify any adverse effect of stratospheric ozone depletion upon skin cancer risk.

UVR and skin cancer risk

UVR exposure was first linked experimentally to skin cancer in the 1920s. Using a mercury-vapour lamp as a source of UVR, Findlay exposed mice experimentally to daily doses of UVR over 58 weeks. Malignant tumours developed in four of the six mice that developed tumours, leading to the conclusion that exposure to UVR could result in skin cancer. Epidemiologists' interest in this association was further stimulated by the possibility of human-induced damage to stratospheric ozone, first theorised in the 1970s. The International Agency for Research on Cancer in 1992 concluded that solar radiation is a cause of skin cancer.

Within the ultraviolet radiation waveband, the highest risk of skin cancer is related to UVB exposure. UVB is much more effective than UVA at causing biological damage, contributing about 80% towards sunburn while UVA contributes the remaining 20%. UVB exposure has been linked conclusively to cutaneous malignant melanoma (CMM) and non-melanoma skin cancer (NMSC). Figure 1shows diagrammatically the UV spectrum and the erythemal effectiveness of solar radiation in humans.

There is a strong relationship between the incidence of all types of skin cancer and latitude, at least within homogeneous populations. Latitude approximately reflects the amount of UVR reaching the earth's surface.

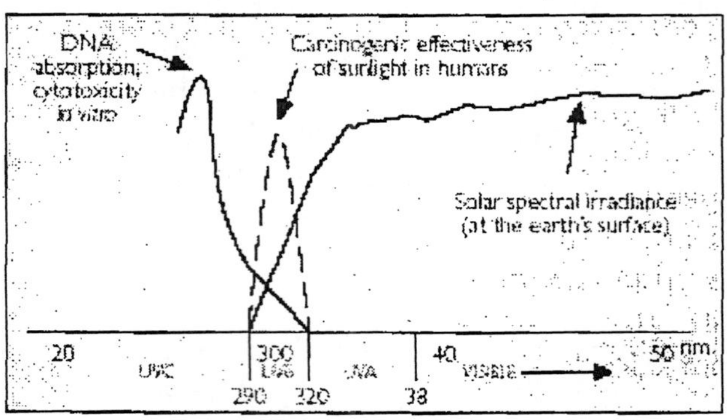

Figure 1 Biologically active UV radiation.

This is due partly to the differing thickness of the ozone layer at different latitudes, and partly to the angle at which solar radiation passes through the atmosphere. In response to UVB exposure the epidermis thickens via an increase in the number of cell layers. This occurs particularly in people who do not tan readily. This thickening reduces the amount of UVB penetration to the basal layer providing partial natural protection against the harmful effects of UVR. Animal experiments indicate that despite this epidermal protection, further UVB exposure can act as a potent tumour promoter on damaged basal cells.

Skin cancer and ozone depletion

Scientists expect the combined effect of recent stratospheric ozone depletion, and its continuation over the next one to two decades, to be an increase in skin cancer incidence in fair-skinned populations living at mid to high latitudes. Future impacts of ozone depletion on skin cancer incidence in European and North American populations have been modelled. The first entails no restrictions on CFC emissions. The second, reflecting the original Montreal protocol of 1987, entails a 50% reduction in the production of the five most important ozonedestroying chemicals by the end of 1999. In the third scenario, under the Copenhagen amendments to that protocol, the production of 21 ozone-depleting chemicals is reduced to zero by the end of 1995. This modelling study estimated that, for the third scenario, by 2050 there would be a park relative increase in total skin cancer incidence of 5-10% in "European" populations living between 40°N and 52°N.

It must be remembered that all such modelling makes simplifying assumptions and entails a substantial range of uncertainty. Not only is the shape of the UVR cancer relationship poorly described in human populations, but also there is inevitable uncertainty about actual future gaseous emissions; the physical interaction between human-induced disturbances of the lower and middle atmospheres; and future changes in patterns of human exposure-related behaviours.

Disorders in Eyes

Both age related macular degeneration (AMD) and cataract show associations with low or depleted antioxidant status and higher oxidative stress, suggesting common aetiological factors. Approximately 50% of incident UVA and 3% of UVB penetrates the cornea, where a further 1% of UVB is absorbed by the aqueous humour. Remaining UVR is absorbed by the lens, hence the UVR association with lens opacities is the most plausible. There is some evidence that sunlight exposure may be implicated in macular degeneration.

Solar radiation and risk of lens opacities

The shorter wavelength constituents of solar radiation are more damaging to biological molecules than is visible light. Although UVB is only 3% of the UVR that reaches the earth, it is much more biologically active than UVA.

In vivo and in vitro laboratory studies demonstrate that exposure to UVR, in particular to UVB, in various mammalian species induces lens opacification. The actual mechanisms remain unclear but a range of adverse effects is observed as a result of free radical generation from UVR energised electrons. There has been criticism that UVR doses in laboratory studies are much higher than those encountered in natural conditions. However, based on ambient UVA and UVB fluxes in the northeastern United States, it has been estimated that 26 hours of continuous UVA exposure or 245 hours of continuous UVB exposures at those ambient levels would exceed the rabbit lens threshold for lens damage. While direct extrapolation from animal studies to humans is not possible it is plausible that in humans, with much longer age spans than laboratory animals, cumulative damage to the lens from UVR could explain the high prevalence of lens opacities in elderly people.

There is mixed evidence for UVR's role in lens opacities in human populations. Cataracts are more common in some countries with high UVR levels. However, few studies have examined whether UVR can explain differences between populations in the prevalence of lens opacities. One study of cataract surgical rates in the United States' Medicare programme estimated a 3% increase in the occurrence of cataract surgery for each 1° decrease in latitude across the United States. However, surgery rates are not a good measure of the prevalence of opacities in the population; they are influenced by service access and differences in the thresholds for eligibility for surgery.

Studies measuring UVR or outdoor exposure in individuals have shown inconsistent results. The strongest evidence is provided by a study of a high UVR exposed group, in the Chesapeake Bay Watermen Study in the United States, which showed an association between adult UVR dose and risk of cortical and posterior subcapsular opacities.

In general population studies in the United States, UVR exposure was related to cortical opacities in one study but not in another, or has been observed in men but not in women. Further support for the association with cortical opacities and UVR comes from mannikin studies showing the largest doses of UVR to be received by the lower and inner (nasal) lens-the site where cortical opacities predominate. Cortical opacities are rare in the upper lens. However, it has been suggested that the lack of an association between UVR and nuclear opacities may reflect failure to measure exposures occurring in earlier life. Since the nuclear material is the oldest in the lens capsule, the most relevant exposures are those that occur in early life. In India, where rates of lens opacities are higher than in Western populations, estimated lifetime sunlight exposure was associated with all types of lens opacities, including nuclear.

An evaluation of the possible risk from UVR must take account of both confounding factors and factors that may modify the association. Factors that may increase susceptibility to UVR-induced damage include poor nutrition and smoking. Smoking may act as an additional source of oxidative stress and consistently has been shown to increase the risk of cataract. Antioxidant micronutrients may enhance the free radical scavenging defence system of the eye. There is some evidence that low dietary intakes of vitamins C, E and carotenoids increase cataract risk.

Effects on the Cornea and Conjunctiva

Acute exposure of the eye to high levels of UVR, particularly in settings of high light reflectance such as snow-covered surroundings, can cause painful inflammation of the cornea or conjunctiva. Commonly called snow-blindness, photokeratitis and photoconjunctivitis are the ocular equivalent of acute sunburn. Pterygium is a common condition that usually affects the nasal conjunctiva, sometimes with extension to the cornea. It is particularly common in populations in areas of high UVR or high exposure to particulate matter. Studies of the Chesapeake Bay watermen showed a dose-response relationship between history of exposure to UVR and risk of pterygium.

Effects on the Retina

Other eye disorders associated with UVR are uncommon

but cause significant morbidity to affected individuals. Acute solar retinopathy, or eclipse retinopathy, usually presents to medical attention soon after a solar eclipse when individuals have looked directly at the sun. Effectively this is a solar burn to the retina. Usually the resulting scotoma resolves but there may be permanent minor field defects. Several cases of solar retinopathy in young adults, possibly related to sun-gazing during a period of low stratospheric ozone in the United States, have been described.

ROLE OF IMMUNE SYSTEM FUNCTION AND IMMUNE-RELATED DISORDERS

Although most of the available evidence comes from studies of experimental animals, it appears that ultraviolet radiation suppresses components of both local and systemic immune functioning. An increase in ultraviolet radiation exposure therefore may increase the occurrence and severity of infectious diseases and, in contrast, reduce the incidence and severity of various autoimmune disorders. The damping down of the T lymphocyte, or " T_H1 ", component of the immune system may alleviate diseases such as multiple sclerosis, rheumatoid arthritis and insulin-dependent diabetes.

Undifferentiated T_H0 cells are immunologically primed to develop into either T_H1 or T_H2 cells; in animals these two groups are thought to be mutually antagonistic. Thus UVR exposure theoretically could worsen T_H2-mediated disease by suppressing T_H1 cell function, however, more recent work has shed some doubt on this notion. In mice UVR exposure is associated with decreased systemic T_H2 as well as T_H1 immune responses. UVR leads to increased secretion of the cytokine, interleukin (IL)-10 appears to suppress T_H1 and T_H2 cytokine responses to external antigens. Much remains unknown. Partly in response to questions about the biological impacts of stratospheric ozone depletion, among scientists there is new interest in assessing the influence of ultraviolet radiation upon immune system function, vitamin D metabolism and the consequences for human disease risks.

Recent research suggests that UVR exposure can weaken TH1-mediated immune responses through several mechanisms:

— UVR can cause local epidermal immunosuppression and a reduction in contact hypersensitivity (CH) and delayed type hypersensitivity (DTH);

— UVR acts to convert urocanic acid (UCA) from the trans-UCA form to its isomer, the cis-UCA form, within the stratum corneum. This process induces changes in epidermal cytokine profiles from a wide range of cell types. UVR-induced DNA damage also alters cytokine profiles, leading to immunosuppression. Liposome therapy with a DNA repair enzyme can prevent UVR-induced cytokine alterations such as the upregulation of IL-10. Importantly, subepidermal cytokine signalling alterations also can induce soluble products that can exert systemic immunosuppression;

— sunlight suppresses secretion of the hormone melatonin. Activation of melatonin receptors on T helper cells appears to enhance T lymphocyte priming and the release of T_H1 type cytokines such as interferon gamma;

— a role for UVR in promoting the secretion of melanocyte stimulating hormone (MSH), which may suppress T_H1 cell activity, also has been proposed;

— the active form of vitamin D, derived from UVR-supported biosynthesis has well-documented immunomodulatory effects. Peripheral monocytes and activated T helper cells have vitamin D receptors, vitamin D or its analogues can down-regulate T helper cell activity.

Overall, these findings indicate that UVR suppresses T_H1-mediated immune activity. It is important to note that part of this effect occurs independently of vitamin D.

Effect on Human Infectious Disease Patterns

Higher UVR exposure could suppress the immune responses to infection of the human host. The total UVR dose required for immune suppression is likely to be less than that required for skin cancer induction but direct human data are not available. In animals, high UVR exposure has been shown to decrease host resistance to viruses such as influenza and cytomegalovirus, parasites such as malaria and other infections such as Listeria monocytogenes and Trichinella spiralis.

Recently, data from these animal studies have been used to develop a model to predict the possible changes in infection patterns in humans due to increased UVR resulting from stratosphere ozone depletion. Importantly, the model did account for likely inter-species variation in susceptibility to UVR-induced immunosuppression. The theoretical model demonstrated that outdoor UVB exposure levels could affect the cellular immune response to the bacteria Listeria monocytogenes in humans. Using a worst-case scenario, ninety minutes of noontime solar exposure in midsummer at 40°N was predicted to lead to a 50% suppression of human host lymphocyte responses against Listeria monocytogenes. A 5% decrease in ozone layer thickness might shorten this exposure time by about 2.5%.

Human epidemiological studies are required to confirm the findings from laboratory or animal studies.

They also are needed to provide clearer risk assessments of the adverse immunosuppressive effect of increased UVR exposure. Personal UVR exposure in humans has been demonstrated to increase the number and severity of orolabial herpes simplex lesions. Recent questions have been raised about the potential adverse consequences of UVR-induced immunosuppression for HIV-infected individuals.

A 1999 review concluded that despite experimental evidence in laboratory animal studies demonstrating HIV viral activation following UV radiation, there were no data in humans that consistently showed clinically significant immunosuppression in HIV-positive patients receiving UVB or PUVA therapy. A small follow-up study of HIV-positive individuals failed to detect any association between sun exposure and HIV disease progression.

Increased UVR Exposure and Effect to Reduce Vaccine Efficacy

There has been concern that increased exposure to UVR due to stratospheric ozone depletion could hamper the effectiveness of vaccines; particularly BCG, measles and hepatitis. BCG vaccine efficacy has a latitudinal gradient with reduced efficacy at lower latitudes. Seasonal differences in vaccine efficacy have been observed for hepatitis B. While this ecological observation may reflect other latitude-related factors, it is also consistent with UVB depressing an effective host response to intradermally administered vaccines. In animal studies, pre-exposure to UVB prior to intradermal vaccination with Mycobacterium bovis (BCG) impairs the DTH immune response of the host animal to mycobacterial antigens. Local UV irradiation of the skin prior to, and following, inoculation decreases the granulomatous reaction to lepromin in sensitised individuals.

Incidence of Non-Hodgkin's Lymphoma

The incidence of Non-Hodgkin's Lymphoma (NHL) has increased greatly worldwide in recent decades. The reasons for this increase are not known but high personal UVR exposure has been suggested as a possible contributary factor, for the following reasons:

— NHL incidence in England and Wales is positively associated with higher solar UV radiation by region;

— patients with NHL also have been noted to have an increased likelihood of non-melanoma skin cancer;

— chronic immunosuppression is an established risk factor for NHL and, as discussed, UVR has immunosuppressive effects on humans.

A causal link has not been established. Nevertheless, NHL is a disease that should be monitored closely because of its possible increase with any future increases in UVR.

UVR Exposure Beneficial for Some Autoimmune Diseases

Recent developments in photoimmunology and epidemiology suggest that UVR may have a beneficial role in autoimmune diseases such as multiple sclerosis (MS), type 1 diabetes mellitus (IDDM) and rheumatoid arthritis (RA). Each of these autoimmune diseases is characterised by a breakdown in immunological self-tolerance that may be initiated by an inducing agent such as an infectious microorganism or a foreign antigen. A cross-reactive autoimmune response occurs and a "self-molecule" is no longer self-tolerated by the immune system. At this stage, the host tissue becomes immunogenic, attracting a T helper cell type 1 (T_H1) mediated immune response resulting in chronic inflammation. That is, the T_H1 lymphocytes no longer recognise the host tissue as such and instead try to eliminate the host tissue by inflammation.

The well-established gradient of MS increasing with increasing latitude may reflect differential UV-induced immune suppression of autoimmune activity. That is, at lower latitudes where MS prevalence is lower high levels of UVR exposure may dampen down the immune overactivity that occurs in MS. In particular, the autoimmune profile of MS is characterised by disturbances of those T cellrelated activities specifically affected by UVB.

A strong inverse association between UVR exposure and MS has been shown.A recent case-control study found that compared to indoor workers living in a low sunlight region, the odds ratios for an outdoor worker dying from MS in low, medium and high residence sunlight were, respectively, 0.89, 0.52 and 0.24. Thus high residential and occupational solar exposure were associated with a reduced likelihood of MS. UVR may affect not only the development of MS but also its clinical course.

For type 1 diabetes, a disease resulting from T cell-mediated inflammation with destruction of pancreatic tissue, the epidemiological evidence also suggests a possible beneficial role for UVR. An increasing disease prevalence gradient with increasing latitude has been noted. In a Finnish birth cohort study, vitamin D supplementation in infancy was inversely associated with subsequent type 1 diabetes. Vitamin D receptor gene allelic status has been found to relate to MS and type 1 diabetes in some populations. For rheumatoid arthritis, dietary supplementation with vitamin D has been related to lower levels of disease activity.

Other Diseases with Immune Dysfunction and UVR

Although the three diseases above are characterised by T_H1 cell overactivity, other immune diseases may be characterised by T_H2 cell overactivity or a mixed T cell overactivity pattern. Systemic lupus erythematosus (SLE) is characterised by a mixed T_H2/T_H1 disturbance. It has been postulated that the immune dysfunction in SLE begins under the skin where UV-induced keratinocytes produce antigens that are recognised by the body to be foreign. UVR plays a major role in the induction of lesions of patients with the cutaneous form of lupus disease and photo-aggravation of systemic disease may occur in systemic SLE.

Atopic eczema, a disease of immune disturbance that includes T_H2 overactivity, appears to be inversely related to UVR. Strong latitudinal gradients for increasing eczema with increasing latitude have been reported in the Northern Hemisphere. In a clinical trial, narrow-band UVB therapy significantly improved allergic eczema. Thus, high UVB exposure appears to have a beneficial effect on the immune disorder of atopic eczema even though this disease is not characterised by a purely T_H1 immune overactivity pattern.

Other Diseases that could be Exacerbated by Decreased UVR Exposure

Certain cancers have been linked to vitamin D deficiency, although not conclusively. Vitamin D deficiency may increase tuberculosis (TB) risk. Evidence suggests that the explanation for this may reflect the immunological modulation caused by vitamin D. Vitamin D activates one group of white blood cells, the monocytes, thereby increasing their capacity to resist cell infection by the mycobacterium. Further, a recent case-control study showed that the combination of vitamin D deficiency and the "high-risk" allele of the vitamin D receptor gene was strongly associated with the occurrence of TB.

During pregnancy, inadequate maternal UVR exposure in the absence of adequate dietary vitamin D sources will lead to low foetal exposure to vitamin D. As vitamin D appears to be important in neural growth this could influence the developing brain of the foetus. In fact, this has been proposed as an explanation for the finding that winter-born babies appear at increased risk of schizophrenia. Furthermore, inadequate UVR exposure usually is associated with reduced visible light and a reduction in photoperiod. This will alter melatonin levels, a hormone important in maintaining the rhythm of wake/ sleep patterns. Changes in photoperiod also have been related to seasonal affective disorder.

4

International Carbon Market

Market based approaches, and especially emissions trading, have been central to the development of the global climate regime to date.

Two aspects of the climate change problem favour the use of market based approaches such as emissions trading as a policy:

— GHGs mix uniformly in the atmosphere so that the location of emission reductions does not matter.

— Lowering the costs of emission reductions is extremely important, given the scale of global reductions likely needed to meet the ultimate objective of the UNFCCC.

Marketbased approaches recognise that con- trolling emission sources, even within the same country or company, can have different costs. Emissions trading provides affected sources with flexibility and a choice of options for meeting their targets cost-effectively. This could entail implementing energy efficiency measures, adopting better control technologies or purchasing "reductions" from a source whose costs of reducing emissions are lower. Emissions trading encourages reductions to take place where they are the least costly, and offers the potential to significantly reduce the overall costs of meeting climate goals.

Emissions trading also supports the adoption of low-carbon technologies. The development, deployment and dissemination of technology are critical to achieving climate goals. Emissions trading can provide a market price incentive for the introduction of technologies that reduce emissions, and offers an important complement to other policies that promote technology development and transfer. The emergence of international and domestic carbon markets in the past few years has mainly resulted from the framework established under the Kyoto Protocol. The Protocol introduced three marketbased mechanisms-

— International emissions trading (IET),

— Joint Implementation (JI) and

— Clean Development Mechanism (CDM).

These mechanisms are also widely credited with helping to create a market value for GHG emission reductions and creating new markets and investment opportunities, even before the Protocol entered into force. This feature has been referred to as the "genius" of Kyoto. The carbon constraints derived from the Protocol have resulted in the establishment of national and international trading schemes for private sector entities, such as the EU Emissions Trading Scheme (EU ETS). Other countries plan to implement such schemes in the coming years and means to link them are being considered. Separate from schemes that directly engage private sector entities, governments of Annex B countries that have ratified the Protocol are also active "buyers" in project based activities under its CDM and JI mechanisms.

Emerging International Carbon Market

The current emerging international carbon market is not one market, but rather a mosaic of markets that includes allowance-based markets from international and domestic emissions trading schemes; credit-based markets related to the project based mechanisms; and voluntary and sub-national trading markets. Many of these policies are at early stages of implementation. The different policy settings creating these markets and the outlook to 2012 are discussed below.

One common element across these differing markets is the unit of measure for the tradable commodity, with one unit being equal to one tonne of greenhouse gases measured in CO_2 equivalent. However, although the unit of measure is the same, the price of a tonne of CO_2 may vary considerably across different markets that are not directly linked, due to differences in supply, demand, compliance requirements and the nature of the commodity being traded.

In markets where there is greater certainty about compliance requirements and allowance allocations, the price of the commodity tends to be higher than for reductions from CDM and JI projects or in voluntary trading schemes.

This is in part because participants in these markets are engaging in forward transactions that entail assuming risks related to project viability and delivery. For example, ERUs from JI projects will not be physically available in the marketplace until at least 2008, and only if the host Party meets the eligibility requirements for trading. While the CDM will generate CERs before 2008, the first CERs were only issued in October 2005, and most transactions will continue to be for forward delivery. As more certainty evolves around JI and the CDM, and in particular the emission reductions associated with these projects, this is likely to be reflected in the price at which they are contracted and/or traded.

The international trading system established by the Protocol will by and large remain dormant until 2008, the earliest date at which the trading of AAUs can occur. Over the next couple of years, the necessary infrastructure to support the market will need to be put into place.

A number of the required bodies have yet to be established and countries must still put in place their national systems to meet the eligibility requirements for trading. The only component of the overall Kyoto "mechanisms" system currently functioning is the CDM which began operation in December 2001. Because the CDM was the first Kyoto mechanism to be activated it has been the primary focus of many governments, industry and non-governmental organisations.

Role of European Union Emissions Trading Scheme

The international carbon market is currently dominated by the EU ETS which began on January 1, 2005. This scheme is mandatory for all 25 EU member states. In a way, it can be seen as 25 fully-linked domestic emissions trading schemes with a common design and central coordination on some key aspects by the European Commission (EC). Countries likely to, or who are expected to, accede into the EU will also be required to implement domestic trading schemes under the EU ETS. The EU ETS could also include European EFTA countries. Norway has developed an emissions trading scheme compatible to the EU ETS and is in discussions with the EU on possible linkages.

In its first phase lasting through 2007, the EU ETS covers six sectors: electricity generation heat and steam production; mineral oil refineries; processing and production of ferrous metals; cement; bricks and ceramics manufacturing; and pulp and paper. Emission allowances are allocated by national governments to energy intensive plants and installations based on a plan approved by the EC. In total, around 12,700 EU installations are to be covered by the scheme. From 2008, the scheme will coincide with the Protocol's first commitment period, and will continue thereafter in five-year intervals.

The EU ETS, as currently defined, is an international emissions trading system but not a Kyoto Protocol trading system. Compliance units are EU allowances (EUAs), not AAUs. Consequently, linkages have to be defined separate from the Protocol to allow the use of international units. Linkages to the Protocol's project based mechanisms (JI and CDM) have already been established; CERs can be used from 2005 and ERUs from 2008.

The EU ETS also provides the possibility of linking with other trading schemes through negotiated agreements. The EU ETS is a policy tool for managing emissions of firms in key industrial sectors. It does not control the possible buying and selling of allowances by individual EU Member States as they manage their compliance for 2008-2102 under the Protocol. The EU ETS is therefore only part of the overall international carbon market emerging in Europe.

Role of Green Investment Schemes

Proposals for "greening AAUs" have emerged as a result of a desire on the part of some Annex I countries to enhance the political acceptability of purchasing AAUs from certain EIT countries when these are seen as deriving from the decline of their economies subsequent to the Kyoto target base year. Although there has been significant interest in GIS, the concept is not yet well-defined and, as yet, there is no real market for greened AAUs.

In simple terms, GIS involves ensuring that revenues from the purchases of surplus AAUs are directed to projects that generate real environmental benefits. Green Investment Schemes are not a defined element of the Kyoto Protocol, but are a preference on the part of buyers and, consequently, there is no formal or widely agreed definition of "green credits."

A distinction is sometimes made between "hard" greening, defined to include only activities or projects that lead to GHG emission reductions that can be directly quantified, and "soft" greening, which includes activities for which GHG emission reductions are not easily quantified. In the "hard" greening scenario, there is a direct relationship between the quantity of emission reductions generated by the activity and the corresponding number of AAUs that are greened. The project based nature of hard greening means that monitoring, reporting and verification processes similar to those introduced for JI or the CDM would be required.

So-called "soft" greening includes a wide range of policy, programme, technology and capacity-building initiatives that are not easily quantified in terms of emission reductions, but which may contribute significantly to action on climate change in the host country. A greened AAU transaction also allows for considerable flexibility in the timing of the exchange of AAUs, the disbursement of funds and the resulting emission reductions. This means that it is possible to target activities that may not necessarily achieve near-term measurable reductions, but which make significant contributions in the medium or longer term.

Role of Joint Implementation (JI)

Trading under JI does not formally begin until 2008, although JI projects could have begun in 2000. Some JI projects are already under development and a forward market for Emission Reduction Units (ERUs) from JI projects is emerging. The main buyers are governments in the EU and Japan; carbon funds such as those under the World Bank; and private entities covered by the EU ETS, which allows ERUs to be used for compliance beginning in 2008.

JI has two "tracks." Where a host country has met all of the requirements for eligibility to participate in international emissions trading, trades under Track 1 JI are possible. Under Track 1, the decision about how many ERUs a project has generated is the responsibility of the countries involved. If only a smaller subset of the eligibility requirements are met, the project has to go through Track 2 JI, a process somewhat akin to the CDM, and the number of ERUs generated will be verified through this process.

Clean Development Mechanism (CDM)

In the three-and-a-half years since the adoption of the Marrakesh Accords in 2001, the majority of experience with the Kyoto mechanisms has been gained in the CDM. This experience relates to project development and the regulatory processes required for registration as CDM project activities, such as the methodology approval process, accreditation of Designated Operational Entities (DOEs) and the project registration process. In October 2005, the final phase of the CDM project cycle, verification of emission reductions and issuance of CERs, became operational.

Important financial flows into host countries are expected to take place as a result of CDM activities; to date more than US$0. million has been allocated to carbon funds or CDM/JI programmes. Together with private and other sources of funding, the OECD conservatively estimates financing for CER purchases under the CDM to 2012 at roughly US$1 billion. The CDM is the most fully developed of the three mechanisms and the most complex.

While the mechanism has made considerable progress since its inception, stakeholders in the CDM have raised concerns over the efficiency and cost of the process. However, given the project-specific nature of the CDM, the necessity to ensure environmental integrity of the system and a steep learning curve by all stakeholders, a rigorous process has been needed. It is also important to note that the Executive Board has instituted measures designed to speed up decisions and reduce their backlog of work.

The CDM has also been burdened by potentially unrealistic expectations about what it can deliver, in terms of the potential size of the market and the ability to bring about major changes in developing countries. Project based mechanisms are, by their nature, more administratively demanding and costly than cap-andtrade. These limitations, in turn, restrict the ability of project based mechanisms to effect the types of infrastructure change and technology shifts that many non-Annex I countries, in particular, had hoped to achieve through the CDM.

In the current carbon market, prices for emission reductions from CDM projects are influenced by expectations about the ability of projects to attain status as CDM project activities and the future delivery of CERs. Because the emission reductions related to the projects are subject to a great deal of risk, namely that they have not undergone registration and verification procedures, these reductions tend to have a low purchase price. Of greater concern, however, is that without a clear signal from policy-makers on post-2012, the CDM will likely experience a loss of interest from carbon investors given the lead time required in developing and implementing a project. As a result, the CDM is expected to begin experiencing a significant slow down of activity in the near future.

Domestic emissions trading systems of various forms are being implemented by a number of countries to help meet their Kyoto commitments. Each of these trading schemes will require decisions on the mechanism, if any, for linking with international and other domestic systems.

— The U.K. launched a GHG emissions trading scheme in 2002 as a main feature of its Climate Change Programme. This scheme had two key elements. The first provided an exemption to the climate change levy (tax) for firms adopting an intensity target. The second was a voluntary trading programme open to all U.K.-based legal entities with direct or indirect GHG emissions not covered by other agreements or directives.

  The targets under this second element were fixed. A unique "gateway" feature connected\the two elements to manage concerns about the linking of fixed and intensity schemes. The U.K. scheme is now going through a rollover process to the EU ETS where the coverage of sources is the same.

— Canada's plans are for an emissions trading programme that will cover GHGs from large industrial emitters responsible for approximately half of national emissions. The system is intensity-based, with an overall goal of improving carbon efficiency by 15 per cent. The government has also committed to provide a price assurance mechanism to ensure that large emitters will be able to meet their regulatory obligations at a cost of no more than CDN$15/tonne for the period 2008-2012. A unique element of the Canadian system is its large provision for domestic offsets.

— While Japan has yet to decide whether to implement emissions trading in its domestic policy mix, both the

Japanese government and its major industries have been significant early participants in the CDM, GIS and JI credits market.

— Switzerland has legislation in place that allows emissions trading by large industries beginning in 2008.

GHG emissions trading systems unrelated to the Protocol are also emerging:

— In the U.S., under a Regional Greenhouse Gas Initiative, 11 northeast states are heading towards an emissions trading system to manage CO_2 emissions in the power sector. There are also markets for carbon offset credits that derive from utility regulatory requirements in some states. The Chicago Climate Exchange (CCX) is a voluntary industry GHG trading pilot programme for emission sources and offset projects in the United States, Canada, Mexico and Brazil.

— In Australia, state governments have recently announced their intention to establish a state level-driven national emissions trading programme covering major power generation and industry sources. New South Wales already has a "baseline and credit" carbon trading market for electricity retailers under the Greenhouse Benchmark Scheme.

— To cater to the interest in "carbon offsets" of a growing number of corporate and government buyers worldwide, a number of international organisations offer carbon offsets from project based forestry and/ or energy initiatives.

International Carbon Market Prospects to 2012

Several key messages can be drawn from the foregoing

discussion on the current and emerging international carbon market:

— Rather than a single carbon market there is a number of markets that differ in timing, location, relationship to the Protocol and their compliance-based versus voluntary nature. While the underlying commodity may seem the same, the buyers, sellers and carbon prices can be quite different.

The dominant market is that created by the EU ETS, which from 2008 is also connected to Kyoto compliance. There is also an active Kyoto compliance market now, mostly involving the governments of Annex B countries through the CDM and JI project based mechanisms

— Whether and how any of these markets begin to merge depends on whether linkages are forged through international and domestic policies. Some connections have occurred through the Kyoto Protocol and the Linking Directive of the EU ETS. But there are gaps; for example, the EU has yet to establish links that allow the private sector entities in the EU ETS to directly participate in international emissions trading, as defined under Article 17 of the Kyoto Protocol.

— Only a relatively small percentage of GHG emission sources in industrialised countries are covered by a trading-based market mechanism linking them to a common international carbon market. While some regional initiatives in the U.S. and Australia are emerging, no U.S. or Australian emissions are currently linked.

The EU ETS and Norwegian emissions trading scheme covers CO_2 only, but not the transport sector or LULUCF. In 2006, the EU will consider whether to examine the scope of the EU ETS to include additional gases and sources. Other Annex B countries have yet to implement emissions trading schemes, although some plan to do so. The opportunity for achieving the potential efficiency and effectiveness attributes of a global trading scheme is, therefore, commensurately limited.

— Capacity-building efforts are likely needed to help some IET countries implement the necessary national inventory systems for eligibility to participate in the Kyoto mechanisms.

— The CDM is a critical first step and is the only Kyoto flexibility mechanism that engages developing countries. While there have been some start-up problems, most of these can be addressed prior to 2012. In the longer term, the CDM is somewhat constrained by its project based framework. A broader scope and/or complementary approaches are likely to be needed for the carbon market to be effective in influencing large-scale capital infrastructure investments in developing countries over the next 20 years.

— Current carbon markets all suffer from the lack of certainty about the role of emissions trading-like market mechanisms in any international climate regime post-2012.

Issues and Options of Beyond—2012

Within a short period of time, the international carbon market has started to take shape. As domestic and international trading schemes continue to come on-line, the market is likely to become more liquid and stable. However, if the carbon market is to play a significant role in helping to achieve the deeper reductions from current emission paths required over the next 20 years, decision-makers will need to consider how best to broaden and deepen the reach of the market. Limitations of emissions trading and project based mechanisms suggest that several key issues will need to be addressed.

1. Countries need to consider how to better engage developing countries in the carbon market in a way that supports the transfer of low-carbon technology and investments in sustainable energy and other sectors. In particular, the engagement of less advanced developing countries in the carbon market needs to be supported. The CDM is currently the only mechanism for developing country participation in the carbon market.

2. The uncertain cost of emissions abatement presents a barrier to both broader participation and deeper reductions. Options to manage cost uncertainty without compromising long-term emission reduction goals should be examined.

3. Domestic policies such as domestic emissions trading systems or crediting mechanisms are needed to enhance the participation of the private sector in the international carbon market. These domestic systems or schemes also determine the extent of coverage of the carbon market and the number of sources that face a common price signal.

   The efficiency of the carbon market and the opportunity to minimise costs depend on linkages between these domestic trading systems. Countries need to consider how best to link systems while addressing national circumstances.

4. In order for the carbon market to impact investment decisions, there must be some assurance that there will

be a value for emission reductions beyond 2012. Since the value of the commodity traded in the international carbon market is entirely based on policies adopted by governments, the market requires a clear signal on the longevity of the limitation and reduction targets by policy-makers. These signals affect not only emissions trading schemes but also project based mechanisms like the CDM. At this stage, aside from the EU's decision to continue the EU ETS beyond 2012, there is little certainty about the path that international climate change policies will take.

Options for the Post-2012 Carbon Market

The issues identified above suggest that there is a need for increased flexibility in the face of cost uncertainties and the requirement for broader participation and greater reductions.A range of options to increase flexibility and address cost uncertainties: dynamic targets; time flexibility; nonbinding targets; price caps; and sectoral approaches. These could modify or complement the current system of targets and mechanisms under the Kyoto Protocol.

Intensity targets

Dynamic or intensity targets are indexed to an agreed variable, such as actual economic growth. Assigned amounts would be adjusted up or down if growth is higher or lower than expected. Although sometimes proposed as a mechanism for accommodating growth, intensity targets and absolute targets can both be defined to address economic growth.

Dynamic targets address uncertainty related to economic growth, but not uncertainty associated with other factors. Whether or not intensity targets are more effective at reducing cost uncertainty than absolute targets is likely to vary among countries depending on the specific relationship between emissions and economic growth. Many of these concerns are particularly relevant to developing countries where wide fluctuations in GDP have occurred, and can be expected to occur, for macroeconomic reasons that do not have commensurate effects on emissions. Although assigned amounts are not known until after the end of the commitment period, dynamic targets can readily accommodate emissions trading.

Significant temporal flexibility

In addition to providing flexibility in the location of emission reductions, emissions trading can also provide significant temporal flexibility. Experience from some emission programmes suggests that flexibility in timing can limit price spikes and reduce cost uncertainty. The Protocol allows emissions to be averaged over a five-year period-essentially allowing banking and borrowing within the first commitment period.

Compliance units can also be banked into future compliance periods. Increased temporal flexibility could be provided by longer compliance periods, but this advantage would need to be weighed against the risk of undermining the environmental objective if sources could defer abatement indefinitely. This risk is more severe in the international arena when participants are sovereign countries and compliance mechanisms correspondingly weaker.

Nonbinding negotiated targets

Emissions do not actually need to be capped for trading to take place. Not all parties need to be potential buyers; some could be only potential sellers to the international carbon market. Potential sellers could adopt voluntary nonbinding negotiated targets. Targets could be set at a national or sectoral level. Credits would be generated and sales would occur only if actual emissions were below the target. The risk of overselling could be limited by making countries responsible for buying back any allowances they had oversold at the end of the commitment period. A commitment period reserve could also be used to reduce the likelihood of overselling.

Nonbinding targets directly address concerns about cost uncertainty and may enable countries to take on more ambitious targets. Nonbinding targets could also provide incentives for developing country participation. In particular, nonbinding country-specific sectoral baselines could be attractive to developing countries seeking to attract major investments in clean technology that fit with their sustainable development priorities, for sectors and sources where a project based mechanism is less applicable.

Safety valve mechanism

A safety valve mechanism involving a maximum price on allowances could limit costs and also cost uncertainty. Two approaches have been proposed:

— economic agents buy allowances at a fixed maximum price from an international body; or

— economic agents within countries buy price-cap allowances from their own governments.

Issues and implications for emissions trading include the process for setting the price cap, how it links with regimes with different price caps and the disposition of revenues if an international body sells price cap allowances. Revenues could be used to fund adaptation or technology research and development.

Effect of Sectoral approaches

Sectoral approaches are the focus of growing international

interest. While broad coverage is important to maximise the efficiency of an emissions trading system, there are a number of reasons why an approach that focuses on a limited number of sectors may be appropriate for inclusion in a future climate regime. From a policy, institutional and economic standpoint, it may be more practical for many countries to start on a sectoral basis than through a national approach.

Only a few key sectors account for the majority of emissions in many countries. Building technical capacity and developing and collecting the necessary data may be much more manageable at a sectoral level. At the same time, adoption of a sectoral approach could support the broader enhancement of emissions monitoring and reporting systems in developing countries.

Sectoral approaches could be fixed or dynamic, and binding or nonbinding. Examples include:

— Sectoral policy-based crediting would generate credits for adopting and implementing climate friendly policies in particular sectors. As already mentioned, a first step toward policy-based crediting has already been taken under the CDM. A project involving the introduction of energy efficiency standards in Ghana is under review. If successful, this could lead to a broad range of policy-based CDM projects. This approach also addresses concerns about possible policy disincentives created by the CDM.

— Country-specific dynamic sectoral crediting baselines could allow developing countries to focus attention on key sectors where investment is in tandem with priorities. This focus also extends to the monitoring and inventory systems needed to cover full sectors rather than individual projects, as with the CDM.

— Transnational sectoral targets could be developed for energy-intensive industries subject to international competition. This approach could be seen as a more effective way of controlling emissions of energy-intensive countries or countries with energy-intensive sectors while addressing competitiveness and leakage concerns. Transnational sectoral targets could coexist with countrywide targets for some or most industrialised countries, in which case transnational obligations would substitute for allocation of national allowances to companies in that sector.

As noted above, these options are likely to complement or modify the existing Kyoto mechanisms. No matter how comprehensive the coverage of targets is in a future regime, there will likely be some countries, sectors and/or sources that are not covered but would be amenable to a project based mechanism. The dual purpose of the CDM will likely continue to be very important as well, ensuring that CDM projects contribute to host country sustainable development. There may also be opportunities to expand the scope of the CDM. One option, policy-based crediting, is already being explored.

If sectoral crediting baselines are included in a future climate regime, consideration will need to be given to whether the CDM or a new process is the appropriate vehicle for implementation. Another approach to broadening the CDM would be to allow wider scope for the inclusion of land use and land-use change activities; this could enable a larger number of countries to participate more actively in the CDM.

There may also continue to be a significant role for a second track JI-like mechanism post-2012 if some countries are unable to meet emissions trading eligibility requirements. The opportunity to link purchases to specific reduction or removal projects may also continue to be important to some buyers, especially if the domestic emissions trading system in the seller country is small or nonexistent.

Global Emissions Trading Systems

The development of a global emissions trading regime could result from future agreements within the UNFCCC or from the "bottom-up" linkage of several trading schemes, or a combination of the two. In general, a globally linked system of markets is desirable because it creates a larger, more liquid market and so should generate bigger cost savings.

The Kyoto Protocol's national emission limits, emissions trading and project based mechanisms provide a direct and relatively straightforward option for linking domestic systems, but other bilateral and multilateral approaches are also possible both within and outside the framework established by the Protocol. The EU Linking Directive has already established linkages between the EU ETS and CDM/JI and the possibility exists that further linkages may be established.

A future global climate regime may include a variety of different types of national and sectoral targets and crediting mechanisms, and this raises questions about the feasibility and desirability of linking different types of trading systems. Merging or linking systems with different types of targets is technically possible from an overall economic perspective, although there can be implications for output, overall emissions and thus, potentially, for environmental integrity. Difficulties can arise when linking an absolute and a rate-based permit trading regime, and also when linking trading regimes that have different monitoring, accounting and enforcement systems.

Elements of a Framework

In order to have a robust and efficient post-2012 international carbon market that is able to deliver on world leaders' expectations and drive investment decisions, there must be demand. Fundamentally, this requires quantitative emission limits, in one form or another.

Related, and equally critical, is the role of participation. The greater the participation of countries and/or coverage of key sources and sectors, the more likely the framework will address competitiveness and leakage concerns. These concerns have contributed to the limited participation of countries in the Protocol's first commitment period. Participation also affects other important attributes of a successful climate regime: environmental effectiveness, cost-effectiveness and political acceptability.

The development of a wider array of options could allow different countries to adopt different policies to address their national circumstances. This could lead to a flexible international framework with a variety of elements linked through an international carbon market.

At an international level, elements in such a framework might include, for example:

— binding fixed emission limits for industrialised countries;

— for industrialised countries unable to agree to the above, binding fixed or dynamic emission limits for some sectors in some regional groupings-or possibly economy-wide binding dynamic emission limits;

— binding transnational sectoral emission limits (fixed or dynamic) for some key sectors represented by multinational "operators" such as cement, steel and aluminum;

— individually customised, voluntary, nonbinding baselines for specific sectors to generate credits, while attracting investment in key sectors of developing countries;

— a project based crediting mechanism to provide coverage of emission reduction and sink enhancement activities not already covered by other marketbased mechanisms.

These options are complementary rather than exclusive. They could be simultaneously implemented, with different countries selecting different policies according to their national circumstances. This would acknowledge a main lesson from the Kyoto process that countries around the world differ widely and may need different forms of commitments.

While an international framework is essential, the ability of the international carbon market to minimise costs and mobilise investment will also depend on private entity participation. Countries will need to consider how best to engage the private sector in the carbon market.

Need for an Early Signal

In order for the international carbon market to have an impact on investment decisions related to long-lived capital stock-energy and transportation systems-the price signal must extend beyond a few years. In particular, the flow of investments through the CDM is likely to diminish significantly over the next couple of years unless there is a clear signal that emission reductions will have value after 2012.

Clearly, it is difficult to provide much assurance at this stage concerning the form of any post-2012 climate regime. Perhaps more important, individual countries that implement domestic emissions trading systems can send a clear signal through policy or legislation that the systems will continue beyond 2012. The EU has already signalled that its ETS will continue beyond 2012 and Canada has indicated its intent to establish longer-term targets for large emitters beyond 2012; other countries could follow.

5

Global Warming

Ever since the invention of the thermometer, some amateur and professional scientists had recorded the temperature wherever they happened to be living or visiting. During the 19th century, government weather services began to record measurements more systematically. By the 1930s, observers had accumulated millions of numbers for temperatures at stations around the world. It was an endlessly challenging task to weed out the unreliable data, average the rest in clever combinations, and compare the results with other weather features such as droughts.

Many of the players in this game pursued a hope of discovering cycles of weather that could lead to predictions. Adding interest to the game was a suspicion that temperatures had generally increased since the late 19th century—at least in eastern North America and western Europe, the only parts of the world where reliable measurements went back so far.

In the 1930s, the press began to call attention to numerous anecdotes of above-normal temperatures. The head of the U.S. Weather Bureau's Division of Climate and Crop Weather responded in 1934. "With 'Grand-Dad' insisting that the winters were colder and the snows deeper when he was a lad," he said, "...it was decided to make a rather exhaustive study of the question." Averaging results from many stations in the eastern United States and some scattered locations elsewhere around the world, the weather services found that 'Grand-Dad' was right: there had indeed been a rise of several degrees Fahrenheit (°F) since 1865 in most regions. Experts thought this was simply one phase of a cycle of rising and falling temperatures that probably ambled along for centuries.

As one scientist explained, when he spoke of the current "climate change" he did not mean any permanent shift, but a long-term cyclical change "like all other climate fluctuations." It may have been the press reports of warming that stimulated an English engineer, Guy Stewart Callendar, to take up climate study as an amateur enthusiast. He studied global temperature change in a systematic and thorough fashion, the first person ever to do so. If anyone else had thought about it, they had presumably been discouraged by the scattered and irregular character of the weather records plus the common assumption that average climate scarcely changed over the span of a century.

After countless hours of sorting out data and penciling sums, Callendar announced that the temperature had definitely risen between 1890 and 1935, all around the world, by close to half a degree Celsius (0.5°C, equal to 0.9°F). Callendar's statistics gave him confidence to push ahead with another and more audacious claim. Reviving an old theory that human emissions of carbon dioxide gas (CO) from burning fuel could cause a "greenhouse effect," Callendar said this was the cause of the warming. It all sounded dubious to most meteorologists.

Temperature data were such a mess of random fluctuations that with enough manipulation you could derive all sorts of spurious trends. Taking a broader look, experts believed that climate was comfortably uniform. "There is no scientific reason to believe that climate will change radically in the next few decades," the highly respected climatologist Helmut Landsberg explained in 1946. "Good and poor years will occur with approximately the same frequency as heretofore." If during some decades there was an unmistakable climate change in some region, that must be just a portion of some local cycle, and in due time the climate of the region would revert to its average.

By the end of the 20th century, scientists were able to check Callendar's figures. They had done far more extensive and sophisticated analysis of the weather records, confirmed by "proxy" data such as studies of tree rings and measurements of old temperatures that lingered in deep boreholes.

The data showed that the world had in fact been warming from the mid 19th century up to about 1940, mostly because of natural fluctuations. As it happened, most of the warming had come in the relatively small patch of the planet that contained the United States and Europe —and thus contained the great majority of scientists and of those who paid attention to scientists. But for this accident, it is not likely that people would have paid attention to the idea of global warming for another generation.

During the 1940s only a few people looked into the question of warming. A prominent example was the Swedish scientist Hans Ahlmann, who voiced concern about the strong warming seen in some northern regions since early in the century. But in 1952, he reported that northern temperatures had begun to fall again since around 1940. The argument for warming caused by CO emissions, another eminent climatologist wrote in 1949, "has rather broken down in the last few years" when temperatures in some regions fell. In any case, as yet another authority remarked, compared with the vast slow swings of ice ages, "the recent oscillations of climate have been relatively small."

If the North Atlantic region was no longer warming, through the 1940s and 1950s it remained balmy in comparison with earlier decades. Increasingly, doubts were voiced about the general assumption of climate stability. Several scientists published analyses of weather records that confirmed Callendar's finding of an overall rise since the 1880s. An example was a careful study of U.S. Weather Bureau data by Landsberg, who was now the Bureau's chief climatologist.

The results persuaded him to abandon his belief that the climate was unchanging. He found an undeniable and significant warming in the first half of the century, especially in more northern latitudes. He thought it might be due either to variations in the Sun's energy or to the rise of CO. Others pitched in with reports of effects plain enough to persuade attentive members of the public. Ahlmann for one announced that glaciers were retreating, crops were growing farther north, and the like.

The respected climate historian Hubert H. Lamb wrote in 1959 that "Our attitude to climatic 'normals' must clearly change," Recent decades could not be called normal by any standard of the past, and he saw no reason to expect the next decades would be "normal" either.

Actually, since the 1930s the temperatures in his own homeland, Britain, had been heading down, but Lamb would not speculate whether that was the start of a cyclical downtrend. It could be "merely another wobble" in one region. Lamb's main point, reinforced by his scholarly studies of weather reports clear back to medieval times, was that regional climate change could be serious and long-lasting. Most meteorologists nevertheless stuck to their belief that the only changes to be expected were moderate swings in one part of the world or another, with a fairly prompt return to the long-term average. If there was almost a consensus that at the present time there was a worldwide tendency to warming, the agreement was fragile.

In January 1961, on a snowy and unusually cold day in New York City, J. Murray Mitchell, Jr. of the U.S. Weather Bureau's Office of Climatology told a meeting of meteorologists that the world's temperature was falling. This was the first time anyone had worked through all the exacting calculations, working out average temperatures for most of the globe, to produce plausible results.

Global temperatures had indeed risen until about 1940, Mitchell said, but since then, temperatures had been falling. There was so much random variation from place to place and from year to year that the reversal to cooling had only now become visible.

Acknowledging that the increasing amount of CO in the atmosphere should give a tendency for warming, Mitchell tentatively suggested that the reversal might be partly caused by smoke from volcanic eruptions and perhaps cyclical changes in the Sun. But "such theories appear to be insufficient to account for the recent cooling," and he could only conclude that the downturn was "a curious enigma." He suspected the cooling might be part of a natural "rhythm," a cycle lasting 80 years or so.

The veteran science correspondent Walter Sullivan was at the meeting, and he reported in the *New York Times* that after days of discussion the meteorologists generally agreed on the existence of the cooling trend, but could not agree on a cause for this or any other climate change. "Many schools of thought were represented... and, while the debate remained good-humored, there was energetic dueling with scientific facts." The confused state of climate science was a public embarrassment.

Through the 1960s and into the 1970s, the average global temperature remained relatively cool. Western Europe in particular suffered some of the coldest winters on record. People will always give special attention to the weather that they see when they walk out their doors, and what they saw made them doubt that global warming was at hand. Experts who had come to suspect greenhouse warming now began to have doubts.

Callendar found the turn worrisome, and contacted climate experts to discuss it. Landsberg returned to his earlier view that the climate was probably showing only transient fluctuations, not a rising trend. While pollution and CO might be altering the climate in limited regions, he wrote, "on the global scale natural forces still prevail." He added, however, that "this should not lead to complacency" about the risk of global changes in the distant future.

It had long been recognised that the central parts of cities were distinctly warmer than the surrounding countryside. In urban areas the absorption of solar energy by smog, black roads and roofs, along with direct outpouring of heat from furnaces and other energy sources, created a "heat island" effect, the most striking of all human modifications of local climate. It could be snowing in the suburbs and raining downtown.

Some pushed ahead to suggest that as human civilisation used ever more energy, in a century or so the direct output of heat could be great enough to disturb the entire global climate. If so, that would not happen soon, and for the moment the main consequences were statistical.

Some experts began to ask whether the warming reported for the decades before 1940 had been an illusion. Most temperature measurements had been made in built-up areas. As the cities grew, so did their local heating, which might have given a spurious impression of global warming. Callendar and others replied that they were well aware of urban effects, and took them fully into account in their calculations.

Mitchell in particular agreed that population growth could explain the "record high" temperatures often reported in American cities—but not the warming of remote Arctic regions. Yet the statistical difficulties were so complex that the global warming up to 1940 remained in doubt. Some skeptics continued to argue that the warming was a mere illusion caused by urbanisation.

While neither scientists nor the public could be sure in the 1970s whether the world was warming or cooling, people were increasingly inclined to believe that global climate was on the move, and in no small way. The reassuring assumption of a stable "normal" climate was rarely heard now. In the early 1970s, a series of ruinous droughts and other exceptionally bad spells of weather in various parts of the world provoked warnings that world food stocks might run out.

Responding to public anxieties, in 1973 the Japan Meteorological Agency sent a questionnaire to meteorological services around the world. They found no consensus. Most agencies believed that there was no clear climate trend, but several (including the Japanese themselves) noted a recent cooling in many regions. Many experts thought it likely that the world had entered a long-term cool spell.

Public pressure was urging scientists to declare where the climate was going. But they could not do so without knowing what caused climate changes. Haze in the air from volcanoes might explain some cooling, but not as much as was observed. As for air pollution from human sources, most experts doubted putting out enough to affect global climate. A more acceptable explanation was a traditional one: the Earth was responding to long-term fluctuations in the Sun's output of energy.

An alternative explanation was found in the "Milankovitch" cycles, tens of thousands of years long, that astronomers calculated for minor variations in the Earth's orbit. These variations brought cyclical changes in the amount of sunlight reaching a given latitude on Earth. In 1966, a leading climate expert analysed the cycles starting on the descent into a new ice age.

In the early 1970s, the nature and timing of the cycles as actually reflected in past climate shifts was pinned down by a variety of measurements, and projecting them forward strengthened the prediction. A gradual cooling was astronomically scheduled over the next few thousand years. Unless, that is, something intervened.

It scarcely mattered what the Milankovitch orbital changes might do, wrote Murray Mitchell in 1972, since "man's intervention... would if anything tend to prolong the present interglacial." Human industry would prevent an advance of the ice by blanketing the Earth with CO.

A panel of top experts convened by the National Academy of Sciences in 1975 tentatively agreed with Mitchell. True, in recent years the temperature had been dropping. Nevertheless, they thought CO "could conceivably" bring half a degree of warming by the end of the century. The outspoken geochemist and oceanographer Wallace Broecker went farther. He suspected that there was indeed a natural cycle responsible for the cooling in recent decades, perhaps originating in cyclical changes on the Sun. If so, it was only temporarily canceling the greenhouse warming.

Meanwhile in 1975, two New Zealand scientists reported that while the Northern Hemisphere had been cooling over the past thirty years, their own region, and probably other parts of the Southern Hemisphere, had been warming. There were too few weather stations in the vast unvisited southern oceans to be certain, but other studies tended to confirm it.

The cooling since around 1940 had been observed mainly in northern latitudes. Perhaps the greenhouse warming was counteracted there by cooling from industrial haze? After all, the Northern Hemisphere was home to most of the world's industry. It was also home to most of the world's population, and as usual, people had been most impressed by the weather where they lived. If there had almost been a consensus in the early 1970s that the entire world was cooling, the consensus now broke down.

Science journalists reported that climate scientists were openly divided, and those who expected warming were increasingly numerous. In an attempt to force scientists to agree on a useful answer, in 1977 the U.S. Department of Defense persuaded two dozen of the world's top climate experts to respond to a complicated survey. Their main conclusion was that scientific knowledge was meager and all predictions unreliable.

The panel was nearly equally divided among three opinions: some thought further cooling was likely, others suspected that moderate greenhouse warming would begin fairly soon, and most of the rest expected the climate would stay about the same at least for the next couple of decades. Only a few thought it probable that there would be considerable global warming by the year 2000.

Government officials and scientists needed more definite statements on what was happening to the weather. Thousands of stations around the world were turning out daily numbers, but these represented many different standards and degrees of reliability—a disorderly, almost indigestible mess.

Around 1980 two groups undertook to work through the numbers in all their grubby details, rejecting sets of uncertain data and tidying up the rest. One group was in New York, funded by NASA and led by James Hansen. They understood that the work by Mitchell and others mainly described the Northern Hemisphere, since that was where the great majority of reliable observations lay.

Sorting through the more limited temperature observations from the other half of the world, they got reasonable averages by applying the same mathematical methods that they had used to get average numbers in their computer models of climate. In 1981, the group reported that "the common misconception that the world is cooling is based on Northern Hemisphere experience to 1970." Just around the time that meteorologists had noticed the cooling trend, such as it was, it had apparently reversed. From a low point in the mid 1960s, by 1980 the world had warmed some 0.2°C.

Hansen's group looked into the causes of the fluctuations, and they got a rather good match for the temperature record using volcanic eruptions plus solar variations. Greenhouse warming by CO had not been a major factor (at least, not yet). More sophisticated analyses in the 1990s would eventually confirm these findings. From the 1940s to the early 1960s, the Northern Hemisphere had indeed cooled while temperatures had held roughly steady in the south. This was largely because of normal variations in natural forces, although industrial aerosol pollution had helped. Then the warming had resumed in both hemispheres. The temporary northern cooling had been bad luck for climate science. By feeding skepticism about the greenhouse effect, while provoking some scientists and many journalists to speculate publicly about the coming of a new ice age, the cool spell gave the field a reputation for fecklessness that it would not soon live down.

Any greenhouse warming had been masked by chance fluctuations in solar activity, pulses of volcanic aerosols, and increased haze from pollution. Furthermore, as a few scientists pointed out, the upper layer of the oceans must have been absorbing heat. These effects could only delay atmospheric warming by a few decades, however.

Hansen's group boldly predicted that considering how fast CO was accumulating, by the end of the 20th century "carbon dioxide warming should emerge from the noise level of natural climatic variability." Around the same time, a few other scientists using somewhat different calculations came to the same conclusion—the warming would show itself clearly sometime around 2000.

The second important group analysing global temperatures was the British government's Climatic Research Unit at the University of East Anglia, led by Tom Wigley and Phil Jones. Help in assembling data and funding came from American scientists and agencies.

The British results agreed overall with the NASA group's findings—the world was getting warmer. In 1982, East Anglia confirmed that the cooling that began in the 1940s had turned around by the early 1970s. 1981 was the warmest year in a record that stretched back a century.

Returning to old records, in 1986 the group produced the first truly solid and comprehensive global analysis of average surface temperatures (including the vast ocean regions, which had been neglected by most earlier studies). They found considerable warming from the late 19th century up to 1940, followed by some regional cooling in the Northern Hemisphere but roughly level conditions overall to the mid-1970s. Then the warming had resumed with a vengeance.

The warmest three years in the entire 134-year record had all occurred in the 1980s. Convincing confirmation came from Hansen and a collaborator, who analysed old records using quite different methods from the British, and came up with substantially the same results. It was true: an unprecedented warming was underway, at least 0.5°C in the past century.

Many thousands of people in many countries had spent much of their lives measuring the weather, while thousands more had devoted themselves to organising and administering the programmes, improving the instruments, standardising the data, and maintaining the records in archives. In geophysics not much came easily. One simple sentence might be the distillation of the labors of a multi-generational global community. And it still had to be interpreted. Most experts saw no solid proof that continued warming lay in the future. After all, reliable records covered barely a century and showed large fluctuations.

A new major effort to track global temperature trends, joining the work by groups in New York and East Anglia, was getting underway at NOAA's National Climatic Data Center in Asheville, North Carolina. The Center had been established in 1951 as the National Weather Records Center to handle the digitised data accumulated by the

Weather Bureau and military services since the 1940s. The staff had assembled the world's largest collection of historical weather records. A team led by Thomas Karl tediously reviewed the statistics for the world and especially the United States.

Each of the three groups began to issue annual updates, which were reported prominently in the press. When all the figures were in for 1988, the year proved to be a record-breaker. But in the early 1990s, average global temperatures dipped. Most experts figured this was caused by the huge 1991 Pinatubo volcanic eruption, whose emissions dimmed sunlight around the world.

Once the volcanic aerosols were washed out, the temperature rise resumed. 1995 was the warmest year on record, but that was topped by 1997, and 1998 beat that in turn by a surprisingly large margin. Of course these were global averages of trends that varied from one region to another.

The citizens of the United States, and in particular residents of the East Coast, had not felt the degree of warming that came in some other parts of the world. But for the world as a whole, for the first time most experts now agreed: a serious warming trend was underway. This consensus was sharply attacked by a few scientists. Some pulled up the old argument that temperature readings were biased by the advance of urbanisation.

In fact, around 1990 meticulous re-analysis of old records had squeezed out the urban heat-island bias to the satisfaction of all but the most stubborn critics. Moreover, long-term warming trends showed up in various kinds of physical "proxy" data measured far from cities. To be sure, in urban areas whatever global warming was being caused by the greenhouse effect got a strong addition of heat, so that the combination significantly raised the mortality from heat waves. But the larger global warming trend was no statistical error.

With the urbanisation argument discredited, the skeptics turned to measurements by satellites that monitored the Earth. Since 1979, when the first of these satellites was launched, they had provided the first truly comprehensive set of global temperature data. The instruments did not measure temperatures on the surface, but at middle heights in the atmosphere. At these levels, the data indicated, there had been no rise of temperature, but instead a slight cooling.

The satellites were designed for observing daily weather fluctuations, not the average that represented climate, and it took an extraordinarily complex analysis to get numbers that showed long-term changes. The analysis turned out to have pitfalls. Some argued against the greenhouse skeptics that the satellite data might even show a little warming.

In an attempt to settle the controversy, a panel of the National Academy of Sciences conducted a full-scale review in 1999 and concluded that the satellites seemed to be reliable. The satellite instruments simply were not designed to see the warming that was indeed taking place at the surface. The fact that higher layers of the atmosphere had not noticeably warmed was embarrassing to the scientists who were constructing computer models of climate, for their models showed significant warming there. They suspected the discrepancy could be explained by temporary effects—volcanic eruptions such as Pinatubo, or perhaps the chemical pollution that was depleting the ozone layer. The skeptics persisted. But most scientists concluded that while the computer models were surely imperfect, the satellite data analysis was too ambiguous to pose a serious challenge to the global warming consensus.

By the late 1990s, there were many kinds of indicators of a general warming at ground level. For example, the Northern Hemisphere spring was coming on average a week earlier than in the 1970s. This was confirmed by such diverse measures as earlier dates for bud-break in European botanical gardens, and a decline of Northern Hemisphere snow cover in the spring as measured in satellite pictures. Turning to a more fundamental indicator, the temperature of the upper layer of the oceans—where nearly all the heat entering the climate system was stored—again a serious rise was found in recent decades.

The 1990s were unquestionably the warmest decade since thermometers came into common use, and the trend was accelerating. Most people now took it for granted that the cause was greenhouse warming, but critics pointed out that other things might be responsible. After all, the greenhouse effect could not have been responsible for much of the warming that had come between the 1890s and 1940, when industrial emissions had still been modest. So announcements that a given year was the warmest on record, when the record had started during the 19th-century cold spell, might not mean as much as people supposed.

The cause of the big warming up to 1940 might be long-term cycles in ocean currents, or variations in the Sun's radiation. There were also decades-long fluctuations in the atmosphere-ocean system and in the global pattern of winds, which drove gradual variations in regional weather patterns. These had been suspected since the 1920s, but only started to become clear in the late 1990s. Until these possibilities were sorted out, the cause of the ground-level warming since 1970 would remain controversial. However, "fingerprints" were found that pointed directly to greenhouse warming. One measure was the difference of temperature between night and day.

Tyndall had pointed out more than a century back that basic physics declared that the greenhouse effect would act most effectively at night. Statistics did show that it was especially at night that the world was warmer. No less convincing, Arrhenius at the turn of the century, and everyone since, had calculated that the Arctic would warm more than other parts of the globe.

The effect was glaringly obvious to scientists as they watched trees take over mountain meadows in Sweden and the Arctic Ocean ice pack grow thin. Alaskans and Siberians didn't need statistics to tell them the weather was changing when they saw buildings sag as the permafrost that supported them melted.

Pursuing this in a more sophisticated way, computer models predicted that greenhouse gases would cause a particular pattern of temperature change. It was different from what might be caused by other external influences, such as solar variations. The observed geographical pattern of change did in fact bear a rough resemblance to the computers' greenhouse effect maps. "It is likely that this trend is partially due to human activities," researchers concluded, "although many uncertainties remain." In a 1995 report, the world's leading experts offered this "fingerprint" as strong evidence that greenhouse warming was truly underway.

A minority of experts continued to question that. Perhaps subtle changes involving the Sun, or perhaps something else, had somehow triggered changes in cloud cover or the like to mimic the greenhouse fingerprint? Yet even if that were true, it just went to show how sensitive the climate must be to delicate shifts in the forces at work in the atmosphere.

A variety of new evidence suggested that the recent warming was exceptional even if one looked back many centuries. Beginning in the 1960s, a few historians and meteorologists had labored to discover variations of climate by digging through historical records of events like freezes and storms. For example, had the disastrous harvest of 1788 helped spark the French Revolution? But it was difficult to derive an accurate picture, let alone quantitative data, from old manuscripts. Better results came from physical analysis of tree rings, coral reefs, and other ingenious proxy measures, which produced increasingly reliable numbers.

One important example was a uniquely straightforward method, the measurement of old temperatures directly in boreholes. Data from various locations in Alaska, published in 1986, showed that the top 100 meters of permafrost was anomalously warm compared with deeper layers. The only possible cause was a rise of average air temperature by a few degrees since the last century, with the heat gradually seeping down into the earth.

In a burst of enthusiasm during the 1990s, scientists took the temperature of hundreds of deep boreholes in rock layers around the planet. The averages gave a clear signal of a recent rise in northern regions. A still more important example of the far-flung efforts was a series of heroic expeditions that labored high into the thin air of the Andes and even Tibet, hauling drill rigs onto tropical ice caps. The hardwon data showed again that the warming in the last few decades was greater than anything seen for thousands of years before. Indeed the ice caps themselves, which had endured since the last ice age, were melting away faster than the scientists could measure them.

Three scientists, combining a variety of measures, put estimated temperatures over the past ten centuries into a graph that showed a sharp turn upward since the start of the industrial revolution. The temperatures of the 1990s soared off the chart. Apparently 1998 had been not just the warmest year of the century, but of the millennium. The graph, widely reprinted, was dubbed the "hockey stick."

The upward turn at the end of the "hockey stick" graph matched the recent rise in greenhouse gases. When the curve of 20th-century temperature change was overlaid with curves showing the predictions of various computer models, simulating the effects of the rising greenhouse gases with adjustments for volcanic eruptions and solar variations, the match was close indeed.

Swift Climate Change

The planet's atmosphere was surely so vast and stable that outside forces, ranging from human activity to volcanic eruptions, could have no more than a local and temporary effect. Looking to times long past, scientists recognised that massive ice sheets had once covered a good part of the Northern Hemisphere. But the Ice Age had evidently ended tens of thousands of years ago, and it was an aberration.

During most of the geological record, the Earth seemed to have been bathed in rather uniform warmth. This opinion became so fixed that, as one meteorologist complained, geology textbooks in 1990 were still copying down from their predecessors the venerable tradition that the age of the dinosaurs, and nearly all other past ages, had enjoyed an "equable climate."

The glacial epoch itself seemed to have been a relatively stable condition that lasted millions of years. It was a surprise when evidence turned up, around the end of the 19th century, that the recent glacial epoch had been made up of several cycles of advance and retreat of ice sheets—not a uniform Ice Age but a series of ice ages. Some geologists denied the whole idea, arguing that every glaciation had been regional, a mere local variation while "the mean climate of the world has been fairly constant." But most accepted the evidence that the Earth's northern latitudes, at least, had repeatedly cooled and warmed as a whole. Global climate could change rapidly—that is, over the course of only a few tens of thousands of years. Probably the ice could come again.

A very few meteorologists speculated about possibilities for more rapid change, perhaps even the sudden onset of an ice age. The Earth's climate system might be in an unstable equilibrium, W.J. Humphreys warned in 1932. Although another ice age might not happen for millions of years, "we are not wholly safe from such a world catastrophe."

The worst scenario was offered by the respected climate expert C.E.P. Brooks. He suggested that a slight change of conditions might set off a self-sustaining shift between climate states. Suppose, he said, some random decrease of snow cover in northern latitudes exposed dark ground. Then the ground would absorb more sunlight, which would warm the air, which would melt still more snow: a vicious feedback cycle. An abrupt and catastrophic rise of tens of degrees was conceivable, "perhaps in the course of a single season." Run the cycle backward, and an ice age might suddenly descend.

Most scientists dismissed Brooks's speculations as preposterous. Talk of sudden change was liable to remind them of notions popularised by religious fundamentalists, who had confronted the scientific community in open conflict for generations. Believers in the literal truth of the Bible insisted that the Earth was only a few thousand years old, and defended their faith by claiming that ice sheets could form and disintegrate in mere decades. Hadn't mammoths been discovered as intact mummies, evidently frozen in a shockingly abrupt change of climate? Scientists scorned such notions. Among other arguments, they pointed out that ice sheets kilometers thick must require at least several thousand years to build up or melt away.

The conviction that climate changed only slowly was not affected by the detailed climate records that were recovered, with increasing frequency from the 1920s through the 1950s, from layers of silt and clay pulled up from the ocean floor. Analysis showed no changes in less than several thousand years. The scientists failed to notice that most cores drilled from the seabed could not in fact record a rapid change. For in many places the mud was constantly stirred by burrowing worms, or by sea floor currents and slumping, which blurred any abrupt differences between layers.

Lakes and peat bogs retained a more detailed record. Most telling were studies in the 1930s and 1940s of Scandinavian lakes and bogs, using ancient pollen to find what plants had lived in the region when the layers of clay ("varves") were laid down. Major changes in the mix of plants suggested that the last ice age had not ended with a uniformly steady warming, but with some peculiar oscillations of temperature.

The most prominent oscillation—already noticed in glacial moraines in Scandinavia around the turn of the century—had begun with a rise in temperature, named the Allerød warm period. This was followed by a spell of bitterly cold weather, first identified in the 1930s using Swedish data. It was dubbed the "Younger Dryas" period after Dryas octopetala, a graceful but hardy Arctic flower whose pollen gave witness to frigid tundra.

The Younger Dryas cold spell was followed by a more gradual warming, ending at temperatures even higher than the present. In 1955 the timing was pinned down in a study that used a new technique for dating, measuring the radioactive isotope carbon-14. The study revealed that the chief oscillation of temperatures had come around 12,000 years ago.

The changes had been rapid—where "rapid," for climate scientists at mid-century, meant a change that progressed over as little as one or two thousand years. Most scientists believed such a shift had to be a local circumstance, not a worldwide phenomenon. Certainly there was no data to drive them to any other conclusion, for it was impossible to correlate sequences of varves between different continents. That would only become possible when radiocarbon dating overcame the many inaccuracies and uncertainties that beset the technique in its early years.

Even swifter changes could show up in the clay varves derived from the layers in the mud of lake beds laid down each year by the spring runoff. But there were countless ways that the spring floods and even the vegetation recorded in the layers could have changed in ways that had nothing to do with climate—a shift of stream drainages, a forest fire, the arrival of a tribe of farmers who cleared the land. Abrupt changes in varves, peatbeds, and other geological records were easily attributed to such circumstances.

Scientists could win a reputation by unraveling causes of kinks in the data, but for climatology it all looked like nothing but local "noise." Thus it was easy to dismiss the large climate swings that an Arizona astronomer, Andrew Ellicott Douglass, reported from his studies of tree rings recovered from ancient buildings and Sequoias. Other scientists supposed these were at most regional occurrences, and indeed even regional climate changes scarcely seemed to affect the trees that most scientists looked at. It didn't help that Douglass tried to correlate his weather patterns with sunspots, an approach most meteorologists thought hopelessly speculative.

If researchers had found simultaneous changes at widely different locations, they might have detected a broad climate shift. But carbon-14 dating remained fraught with uncertainties, and matching up the chronologies of different places was difficult and controversial. Further, even a massive and global climate change could bring rains in one locale, cold in another, and not much shift of vegetation at all in a third. So each study remained isolated from the others. That was compatible with "the uniformitarian principle."

This geological tenet held that the fundamental forces that molded ice, rock, sea, and air did not vary over time. Some further insisted that nothing could change otherwise than the way things are seen to change in the present. The uniformitarian principle was cherished by geologists as the very foundation of their science, for how could you study anything scientifically unless the rules stayed the same? The idea had become central to their training and theories during a century of disputes.

Scientists had painfully given up traditions that explained certain geological features by Noah's Flood or other one-time supernatural interventions. Although many of the theories of catastrophic geological change were argued on fully scientific grounds, by the end of the nineteenth century scientists had come to lump all such theories with religious dogmatism. The passionate debates between "uniformitarian" and "catastrophist" viewpoints had only partly brought science into conflict with religion, however.

Many pious scientists and rational preachers could agree that everything happened by gradual natural processes in a world governed by a reliable God-given order. Nowadays temperatures apparently could not rise or fall radically in less than millennia, so the uniformitarian principle declared that such changes could not have happened in the past. The principle thus went hand-in-glove with a prevailing "gradualist" approach to all things geological.

Alongside physical arguments that the great masses of ice, rock and water could not be quickly changed, paleontologists subscribed to a neo-Darwinian model of the evolution of species which argued that here too change must be continuous and gradual. All that seemed to apply to climate.

Textbooks pointed out, for example, that there were plausible reasons to believe that tropical rain forests had scarcely changed over millions of years, so the climates that sustained the orchids and parrots must have been equally stable. There was no reason to worry about the fact that old carbon-14 dates were accurate only within about a thousand years plus or minus, so that a faster change could hardly have been detected. If there were unmistakeable fluctuations like the Younger Dryas, presumably those had regional rather than global scope —restricted to the vicinity of the North Atlantic or an even narrower area.

In 1956, a change at the fastest speed that anyone expected was discovered by the carbon-14 expert Hans Suess, studying the shells of plankton embedded in cores of clay pulled from the deep seabed by Columbia University's Lamont Geological Observatory. Suess reported that the last glacial period had ended with a "relatively rapid" rise of temperature—about 1°C (roughly

2°F) per thousand years. The rise looked even more abrupt when David Ericson and collaborators inspected the way fossii foraminifera shells varied from layer to layer in the Lamont cores. They reported a "rather sudden change from more or less stable glacial conditions" about 11,000 years ago, a change from fully glacial conditions to modern warmth within as little as a thousand years. They acknowledged this was "opposed to the usual view of a gradual change."

Indeed Cesare Emiliani, who often disagreed with Lamont scientists, published an argument that the temperature rise of some 8°C had been the expected gradual kind, stretching over some 8,000 years. More was at stake than simple dating. A graduate student in the Lamont group, Wallace Broecker, put a bold idea in his doctoral thesis. Looking at this and other data, he found "a far different picture of glacial oscillations than the usual sinusoidal pattern." Like Brooks, he suggested that "two stable states exist, the glacial state and the interglacial state, and that the system changes quite rapidly from one to the other." This was only one passage in a thick doctoral thesis that few people read, and sounded much like Brooks's speculations on cataclysmic changes, long since dismissed by scientists as altogether implausible. After considerable debate, Emiliani won his point.

The rapid shift that Ericson had reported was not really to be found in the data. Like some other sudden changes reported in natural records, it reflected peculiarities in the method of analysing samples, not the real world itself. Yet mistakes can be valuable, if they set someone like Broecker to thinking about overlooked possibilities.

By 1960, three Lamont scientists—Broecker, Maurice Ewing, and Bruce Heezen—were reporting a variety of evidence, from deep-sea and lake deposits, that a radical global climate shift of as much as 5-10°C had indeed taken place in less than a thousand years. While it would necessarily take many thousands of years to melt the great ice sheets, they had realised that meanwhile the atmosphere and the ocean surface waters, which were less massive, could be fluctuating on their own. Broecker speculated that the climate shifts might reflect some kind of rapid turnover of North Atlantic ocean waters—a natural place for an oceanographer to look.

A few scientists responded with more specific models. Most important was a widely noted paper by Ewing and William Donn, who were "stimulated by the observation that the change in climate which occurred at the close of the [most recent] glacial period was extremely abrupt." Their model proposed ways that feedbacks involving Arctic ice cover could promote change on a surprisingly rapid scale. Following up, J.D. Ives drew on his detailed field studies of Labrador to assert that the topography there could support what he called "instantaneous glacierisation of a large area." By "instantaneous" he meant an advance of ice sheets over the course of a mere few thousand years, which was roughly ten times faster than most scientists had imagined.

Further information came from studies of fossil pollen recovered from layers of peat laid down in bogs. Those undertaking such work had not set out to study the speed of climate change—their inquiry was mostly a routine, plodding counting of hundreds of specks under the microscope, assembling data on vegetation shifts to catalog the way ice sheets came and went. But the carbon-14 dates offered surprises for an attentive eye.

During the 1950s, Immanuel Velikovsky and others had excited the public with popular books describing abrupt and marvelous upheavals in the Earth's history. Every mammoth frozen in permafrost was offered as proof that the world's climate could change catastrophically overnight. Experts were weary of explaining to students and newspaper reporters that the scenarios were sheer fantasy. The battle against Velikovsky and his ilk only reinforced geologists' insistence on the uniformitarian principle, which they took as a denial of any change radically unlike changes seen in the present. Ideas of catastrophic change were also tainted by the way they were used, persistently and increasingly, by zealots who sought "scientific" proof for their fundamentalist interpretation of passages in the Bible.

In the late 1950s, a group in Chicago carried out tabletop "dishpan" experiments using a rotating fluid to simulate the circulation of the atmosphere. They found that a circulation pattern could flip between distinct modes. If the actual atmospheric circulation did that, it would change weather patterns in many regions almost instantly. On a still larger scale, in the early 1960s a few scientists created crude but robust mathematical models which demonstrated that global climate really could change to an enormous extent in a relatively short time, thanks to feedbacks in the amount of snow cover and the like.

Probably it was no coincidence that this new readiness of scientists to consider rapid and disastrous global change spread in the early 1960s. That was exactly when the world public was becoming anxious over the possibility of sudden global catastrophe. Alongside the fantasies of Velikovsky, and increasingly shrill warnings from Bible fundamentalists, there were sober possibilities of disaster brought on by nuclear war, not to mention threats to the entire planet from chemical pollution and other human industrial ills.

Now that theoretical ideas and the general trend of opinion alike made it easier for climate scientists to envision sharp change, they were increasingly able to notice it in their data. Broecker in particular, looking at deep-sea cores, in 1966 pointed to an "abrupt transition between two stable modes of operation of the ocean-atmosphere system," especially a "sharp unidirectional change" around 11,000 years ago. It proved possible to build simple fluid-flow models that showed how such a change could be promoted by a switch in the pattern of ocean currents. Improved deep-sea records, going back hundreds of millennia, brought additional information.

By comparing the irregular curves from a number of cores, Broecker noticed that the general pattern of glacial cycles was not a simple symmetric wave. It looked more like a sawtooth where "gradual glacial buildups over periods averaging 90,000 years in length are terminated by deglaciations completed in less than one tenth this time." The view was supported by data gathered independently at the University of Wisconsin-Madison, where Reid Bryson was already interested in rapid climate changes.

In the late 1950s, supported by an Air Force contract to study weather anomalies, he had been struck by the wide variability of climates as recorded in the varying width of tree rings. And he was familiar with the Chicago "dishpan" experiments that showed how a circulation pattern might change almost instantaneously. Bryson brought together a group to take a new, interdisciplinary look at climate, including even an anthropologist who studied the ancient native American cultures of the Midwest. From bones and pollen they deduced that a disastrous drought had struck the region in the 1200s—the very period when the flourishing towns of the Mound Builders had gone into decline. It was already known that around that time a great drought had ravaged the Anasazi culture in the Southwest.

Variety of historical evidence hinted that the climate shift had been worldwide. And there seemed to have been distinct starting and ending points. By the mid 1960s, Bryson concluded that "climatic changes do not come about by slow, gradual change, but rather by apparently discrete 'jumps' from one [atmospheric] circulation regime to another." Next the Wisconsin team reviewed carbon-14 dates of pollen from around the end of the last ice age. In 1968, they reported there had been a rapid shift around 10,500 years ago, and by "rapid" they meant a change in the mix of tree species within less than a century. That was about as fast as a forest could adjust, so the climate itself could have changed even faster. Perhaps the Younger Dryas was not just a local Scandinavian anomaly.

Bryson and his collaborators were developing a systematic technique for translating their counts of different kinds of pollens into a record of rainfall and temperature. It was a technique "built on a foundation of debatable assumptions," as one reviewer observed, yet still "a major step forward." They produced for the American Midwest the most accurate, detailed, and comprehensive climate record available anywhere. Looking at hundreds of carbon-14 dates spanning the past dozen millennia—dates that improvements had made accurate enough to give a reasonable correlation among widely dispersed sites—they believed they could confirm Bryson's disturbing conclusion.

A group of glacial-epoch experts reached in a meeting at Brown University in 1972, agreed that interglacial periods tended to be short and to end more abruptly than had been supposed. In view of the cooling that had been reported in the Arctic since the 1940s, they suspected near the end of the present interglacial period. The majority concluded that the current warm period might possibly end in rapid cooling within the next few hundred years—"a first order environmental hazard." Bryson, Stephen Schneider, and a few others took the concern to the public. For all anyone could say, the next decade might start a plunge into a cataclysmic freeze, drought, or other change unprecedented in recent memory, but not without precedent in the archeological and geological record.

While Bryson warned that the increasing pollution of the atmosphere would shade the Earth and bring rapid cooling, this was not the only possibility. The growing realisation that small perturbations could trigger sudden climate change also impressed scientists who were growing concerned about the rising level of the greenhouse gas carbon dioxide (CO). Perhaps that might bring serious global warming and other weather changes within as little as a century or two.

As abrupt changes became more credible they were seen in still more kinds of evidence. One example was the shells of beetles, which are abundant in peat bogs, and so remarkably durable that they can be identified even 50,000 years back. Beetles swiftly invade or abandon a region as conditions shift, so the set of species you find gives a sensitive measure of the climate. Russell Coope, studying bog beetles in England, turned up rapid fluctuations from cold to warm and back again, a matter of perhaps 3°C, around 13,000 years ago. It all happened within a thousand years at most, he reported. This singular approach got a skeptical response from other scientists as they pursued the well established study of pollens, for they were accustomed to seeing more gradual transformations of forests and grasslands. The fluctuations in Coope's records were easily dismissed as local peculiarities of English beetles.

The Camp Century cores, too, might tell little about change on a global scale. The data might be sensitive to changes of ice cover in the seas near Greenland, or to a local shift of the ice cap's glacial flow. Other evidence, especially oxygen isotopes in shells from deep-sea cores that reflected conditions in the entire North Atlantic, showed changes only over several thousand years. Nevertheless, as pieces of evidence accumulated, a growing number of scientists found it plausible that the climate over large regions, if not the entire world, had sometimes changed markedly in a thousand years or even less. Perhaps one reason was that the early 1970s meanwhile saw further development of global energy-balance models in which a few simple equations produced radical instability.

In particular, Mikhail Budyko in Leningrad pursued calculations about feedbacks involving ice cover, and suggested that at the rate pumping CO into the atmosphere, the ice covering the Arctic Ocean might melt entirely by 2050.

Conversely, a buildup of snow and ice might reflect enough sunlight to flip the Earth into a glaciated state. These ideas prompted George Kukla and his wife Helena to inspect satellite photos of Arctic snow cover, and they found surprisingly large variations from year to year. If the large buildup seen in 1971 were repeated for only another seven years, the snow and ice would reflect as much sunlight as during a glacial period.

Meanwhile glacier experts developed ingenious models that suggested that global warming might provoke the ice sheets of Antarctica to break up swiftly, shocking the climate system with a huge surge of icewater. Bryson and other scientists worked harder than ever to bring their concerns to the attention of the scientific community and the public. As Broecker put it, any decade now the world could be hit by a severe "climatic surprise." Most scientists spoke more cautiously.

When leading experts had to state a consensus opinion, as in a 1975 National Academy of Sciences report on plans for international cooperation in atmospheric research, they were circumspect. Evaluating past statistics, the panel concluded that predictable influences on climate made for only relatively small changes, which would take centuries or longer to develop. Any big jerks that might matter for current human affairs were likely to be just "noise," the usual irregularities of climate.

The panel agreed that there was a significant "likelihood of a major deterioration of global climate in the years ahead," but they could not say how rapidly that might happen. Scientists of the time disagreed on whether the greatest global risk was cooling by atmospheric pollution or greenhouse effect warming. No doubt the present warm interglacial period would end sooner or later, but that might be thousands of years away. About the only thing the scientists fully agreed on was that they were largely ignorant.

As a landscape that looks smooth from a distance may display jagged gullies when seen through binoculars, so sharper and sharper changes appeared as measuring techniques got better. It was getting easier for scientists to consider such colossal transformations, for uniformitarian thinking was under attack. By the early 1980s, some geologists were stressing the importance of rare events like the enormous floods that had drained temporary lakes during the melting of the continental ice sheets.

In biology, Stephen Jay Gould and a few others were arguing that the evolution of at least some species had proceeded in "punctuated" bursts. Other scientists were offering plausible scenarios of cosmic catastrophes that might happen only once in tens of millions of years.

The evidence of abrupt shifts that turned up in occasional studies may seem strong in retrospect, but at the time it was not particularly convincing. Any single record could be subject to all kinds of accidental errors. The best example was in the best data on climate shifts, the wiggles in measurements from the Camp Century core. These data came from near the bottom of the hole, where the ice layers were squeezed tissue-thin and probably folded and distorted as they flowed over the bedrock.

Many continued to believe that the oceans could only vary gradually over thousands of years, with a thermal inertia that must moderate any climate changes. These scientists should have realised that the top few meters of ocean exchange heat only slowly with the rest. And they should have recalled that at most places in the deep sea, sediments accumulate at only a few centimeters per thousand years, with the churning by burrowing worms blurring any record of change. Ice did not have these problems, so further progress would depend on getting more and better ice cores.

Ice drilling was becoming a little world of its own, inhabited by people of many nations (Dansgaard's "Danish" team spoke eight different languages). Their divergent interests made for long and occasionally painful negotiations. But the trouble of cooperation was worth it for bringing in a variety of expertise, plus (no less essential) a variety of agencies that might grant funds.

The ancient ice that drilling teams hunted was at places barely possible to reach —eventually they penetrated not only the polar ice caps, but mountain icefields from Peru to Tibet—and the teams had to somehow get there with tons of equipment and supplies. The outcome was a series of engineering triumphs, which could turn into maddening fiascos when a costly drill head got irretrievably stuck a mile down. Engineers went back to their drawingboards, team leaders contrived to get more funds, and the work slowly pushed on.

A breakthrough came after the ice drillers went to a second location, a military radar station named "Dye 3" some 1,400 kilometers distant from Camp Century. By 1981, after a decade of tenacious labor and the invention of an ingenious new drill, they had extracted gleaming cylinders of ice ten centimeters in diameter and in total over two kilometers long. Dansgaard's group cut out 67,000 samples, and in each sample analysed the ratios of oxygen isotopes.

The temperature record showed what they called "violent" changes—which corresponded closely to the jumps at Camp Century. Moreover, the most prominent of the changes in their record corresponded to the Younger Dryas oscillation that had been recorded in pollen shifts all over Europe. It showed up in the ice as a swift warming interrupted by "a dramatic cooling of rather short duration, perhaps only a few hundred years."

A particularly good correlation came from a group under Hans Oeschger. An ice drilling pioneer, Oeschger was now measuring oxygen isotopes in glacial-era lake deposits near his home in Bern, Switzerland. That was far indeed from Greenland, but his group found "drastic climatic changes" that neatly matched the ice record. The severe cold spells became known as "Dansgaard-Oeschger events." They seemed to be restricted to the North Atlantic and Europe.

As ice drillers improved their techniques, making ever better measurements along their layered cores, they found a variety of large steps not only in temperature but also in the CO concentration. This was a great surprise to everyone. Since the gas circulates through the atmosphere in a matter of months, the steps seemed to reflect worldwide changes. Other scientists promptly pointed out that the observations might be a mere artifact—the amount of gas absorbed might change with the local temperature in Greenland because of the physical chemistry of ice. It remained clear that something had made spectacular jumps. A variety of other evidence for very abrupt climate changes was accumulating, and some began to entertain the notion of such change on a global scale.

Most of these scientists, after presenting their data, could not resist adding a few suggestive words about possible causes. Dansgaard's group was typical in speculating about "shifts between two different quasi-stationary modes of atmospheric circulation." That was the most common idea about how climate might change rapidly, harking back to the "dishpan" experiments of the 1950s. It implied transient variations of wind patterns within broad limits, and mostly concerned how weather might change in a particular region.

The new thinking about grand global shifts urged a broader view. It was hard to see how the atmosphere could settle into an entirely new state unless something drastic happened in the oceans. For it is sea water, not air, that holds most of the heat energy and most of the moisture and CO of the climate system. The question of century-scale shifts, now a main topic in climatology, came to rest on the desks of ocean scientists. Their response was prompt. Experts mooted various hypotheses about how changes in the surface waters might affect CO levels. There were complex linkages among temperature, sea water chemistry, biological activity, and the chemical nutrients that currents brought to the surface. There were also reasons to believe that the pattern of North Atlantic Ocean circulation could change on a short timescale. Since the circulating waters carry tremendous quantities of heat northward from the tropics, if the circulation ground to a halt, temperatures in many regions of the Northern Hemisphere would immediately plunge.

Broecker began to warn that the ocean-atmosphere climate system did not necessarily respond smoothly when it was pushed—it might jerk. In 1987, he wrote that scientists had been "lulled into complacency." People were increasingly taking their cue from elaborate super computer simulations of the general circulation of the atmosphere, failing to realise that the models, in the very way they were constructed, allowed only smooth and gradual changes.

An "unstable" model would have been reworked by its authors until it produced more consistent results. Broecker strongly suspected that "changes in climate come in leaps rather than gradually," posing a drastic threat to human society and the natural world. And indeed new computer models, labouring to incorporate interactions between air and sea, hinted that he was right.

Early in the 1990s, further revelations startled climate scientists. The quantity, variety, and accuracy of measurements of ancient climates was increasing at a breakneck pace—compared with the data available in the 1970s there were orders of magnitude more now in hand.

The first shock came from the summit of the Greenland ice plateau, a white wasteland so high that altitude sickness was a problem. From this location all ice flowed outward, so it was hoped that even at the bottom, three kilometers deep, the layers would be relatively undisturbed by the movement.

Early hopes for a new cooperative programme joining Americans and Europeans had broken down, and each team drilled its own hole, some three kilometers deep. Competition was transmuted into cooperation by a decision to put the two boreholes just far enough apart so that anything that showed up in both cores must represent a real climate effect, not an accident due to bedrock conditions. The match turned out to be remarkably exact for most of the way down. The comparison between cores showed convincingly that climate could change more rapidly than almost any scientist had imagined.

Swings of temperature that were believed in the 1950s to take tens of thousands of years, in the 1970s to take thousands of years, and in the 1980s to take hundreds of years, were now found to take only decades. Ice core analysis by Dansgaard's group, confirmed by the Americans, showed rapid oscillations of temperature repeatedly at irregular intervals throughout the last glacial period. Greenland had sometimes warmed a shocking 7°C within a span of less than 50 years.

In the late 1980s and early 1990s, studies of pollen and the like at locations ranging from Ohio to Japan to Tierra del Fuego, dated with carbon-14 using improved techniques, suggested that the Younger Dryas events had affected climates around the world. The extent of this perturbation, and just how weather had changed in different regions, was controversial. But scientists were increasingly persuaded that abrupt climate shifts could have global scope.

New studies made it plausible that warming of the oceans could cause some of the deposits to disintegrate in a landslide-like chain reaction, venting enough methane and CO into the atmosphere to redouble global warming. The idea sounded like science fiction, and it seemed highly unlikely to happen anytime soon.

Yet in the 1990s, geologists found that just such titanic outbursts had actually caused a spurt of warming 55 million years ago—certainly something back then had radically changed climate, bringing mass extinctions and a new geological era, and clathrates were the leading suspect. The overall rise in temperature back then had apparently stretched over tens of thousands of years, "rapid" only to a geologist. But it seemed to have come in abrupt steps, which in some centuries might have pumped greenhouse gases into the atmosphere at a rate fast enough to bring serious changes within a human lifetime.

Ominously, data showed that sudden climate shifts did not happen only during a glacial period. In 1993, Dansgaard and his colleagues reported that rapid oscillations had been common during the last interglacial warm period—enormous spikes of cooling, like a 14-degree cold snap that had struck in the span of a decade and lasted 70 years. The instability was unlike anything the ice record showed for current interglacial period. The announcement, Science magazine reported, "shattered" the standard picture of benign, equable interglacials. Others soon showed that these measurements, made near the bottom of the core, had been distorted by ice flow that stirred together layers from warm and cold periods. Interglacials were perhaps not so horrendously variable.

But in terms of how scientists thought about the present climate system, one might say that the ice had been broken. People recalled that the present system was certainly subject to abrupt but harrowing droughts, like the one discovered by Bryson that had devastated native North American cultures. Persuasive new geological evidence blamed extreme prolonged droughts for the downfall of ancient Mayan and Mesopotamian civilisations too.

An entirely different sort of evidence for rapid change came from improved observations of Arctic and Antarctic regions. New views from satellites, plus vigorous programmes of precise measurements from airplanes and on the ground, showed that vast glaciers might change their speed of travel and enormous ice sheets might break up entirely within a matter of months.

As one expert remarked, that "ran counter to much of the accepted wisdom regarding ice sheets, which, lacking modern observational capabilities, was largely based on 'steady-state' assumptions." Thus added to all the other feedbacks was a plausible possibility that climate could be transformed by swift alterations of land or sea ice.

The new view of climate was reinforced by one of the last great achievements of the Soviet Union, an ice core drilled with French collaboration at Vostok in Antarctica. The record reached back over nearly four complete glacial-interglacial cycles—and almost every stretch was peppered with drastic temperature changes.

The Antarctic record was too fuzzy to say whether any of these had come and gone on the decade-size timescale of the Younger Dryas. But warm interglacial periods had certainly been subject to big swings of temperature lasting for centuries. Especially striking to the researchers, by contrast, the ten thousand years since the last glaciation. When Bryson, Schneider, and others had warned that the century or so of stability in recent memory did not reflect "normal" long-term variations, they had touched on an instability grander than they guessed. The entire rise of human civilisation since the end of the Younger Dryas had taken place during a period of warm, stable climate that was unique in the long record. The climate known to history was a lucky anomaly.

The accumulation of evidence, reinforced by at least one reasonable explanation destroyed long-held assumptions. Most experts now accepted that abrupt climate change, huge change, global change, was possible at any time.

A report written by a National Academy of Sciences committee in 2001 said that the recognition, during the 1990s, of the possibility of abrupt global climate change constituted no less than a fundamental reorientation of thinking, a "paradigm shift for the research community." The first strong consensus statement had been issued in 1995 by the Intergovernmental Panel on Climate Change, representing the considered views of nearly all the world's climate scientists.

The report included a notice that climate "surprises" were possible—"Future unexpected, large, and rapid climate system changes." The point was not emphasised by the report's authors, and it was seldom mentioned by the press. Despite the profound implications of this new viewpoint, hardly anyone rose to dispute it. Yet while they did not deny the facts head-on, many denied them more subtly, by failing to revise their accustomed ways of thinking about climate. For example, few of the scientists studying pollen in bogs went back to their data and took on the difficult task of looking for catastrophically rapid shifts in the past.

"Geoscientists are just beginning to accept and adapt to the new paradigm of highly variable climate systems," said the Academy committee in 2001. And beyond geoscientists, "this new paradigm has not yet penetrated the impacts community," the economists and other specialists who tried to calculate the consequences of climate change. Policy-makers and the public lagged even farther behind in grasping what the new scientific view could mean.

The actual history shows that even the best scientific data are never that definitive. People can see only what they find believable. Over the decades, many scientists who looked at tree rings, varves, ice layers, and so forth had held evidence of rapid climate shifts before their eyes. They easily dismissed it. There were plausible reasons to believe that global cataclysm was a fantasy of crackpots and Bible fundamentalists.

Records of the past were mostly too fuzzy to show rapid changes, and where such a change did plainly appear, it was readily attributed to something other than climate. Sometimes the scientists' assumptions were actually built into their procedures. When pollen specialists routinely analysed their clay cores in 10-centimeter slices, they could not possibly see changes that took place within a centimeter's worth of layers. If the conventional beliefs had been the same in 1993 as in 1953—that significant climate change always takes many thousands of years—the decade-scale fluctuations in ice cores would have been passed over as meaningless noise.

6

Sea Level Rise

"Sea-Level Rise and Global Climate Change" is the fourth in a series of reports examining the potential impacts of climate change on the U.S. environment and society. The vulnerability of a coastal area to sea-level rise varies according to the physical characteristics of the coastline, the population size and amount of development, and the responsiveness of land-use and infrastructure planning at the local level. Low-lying developed areas in the Gulf Coast, the South, and the mid-Atlantic regions are especially at risk from sea-level rise.

The rapid growth of coastal areas in the last few decades has resulted in larger populations and more valuable coastal property being at risk from sea-level rise. This growth, which is expected to continue, brings with it a greater likelihood of increased property damage in coastal areas. The major physical impacts of a rise in sea level include erosion of beaches, inundation of deltas as well as flooding and loss of many marshes and wetlands. Increased salinity will likely become a problem in coastal aquifers and estuarine systems as a result of saltwater intrusion. Although there is some uncertainty about the effect of climate change on storms and hurricanes, increases in the intensity or frequency or changes in the paths of these storms could increase storm damage in coastal areas. Damage to and loss of coastal areas would jeopardize the economic and ecological amenities provided by coastal wetlands and marshes, including flood control, critical ecological habitat, and water purification. Damages and economic losses could be reduced if local decision-makers understand the potential impacts of sea-level rise and use this information for planning.

Factors Affecting the Vulnerability of the U.S. Coastal Zone

The major coastal regions of the United States differ in their vulnerability to the risks of sea -level rise. Important local and regional factors that affect vulnerability include variations in the physical characteristics of the coastal area, rates of projected population growth and investment, and management policies and practices. These differences will in turn influence the extent of impacts of sea-level rise on coastal areas. Major physical impacts of sea-level rise include the following:

— erosion of beaches, bay shores, and tidily influenced river deltas;

— permanent inundation or wetland colonisation of low-lying uplands;

— increased flooding and erosion of marshes, wetlands and tidal flats, potentially resulting in net degradation and losses as a result of normal tidal inundation and episodic storm surges ;

— increased flooding and storm damage in low-lying coastal areas as episodic storm surges and destructive waves penetrate further inland; and

— increased salinity in estuaries, marshes, coastal rivers, and coastal aquifers.

These primary impacts will trigger other impacts such as damage to buildings and other coastal infrastructure, including ports, ship channels, and bridges. Where hazardous waste landfills are affected, pollutants in the landfills may migrate because of flooding and water-table changes. As sea-level rise accelerates, these impacts may become more severe, depending on individual site characteristics and protection strategies.

Geographic Factors

From a physical standpoint, the East and Gulf coasts are more vulnerable to sealevel rise than the West Coast because the former have extensive low -lying coastal plains, while much of the latter is composed of cliffs Overall, the coastal zones of the Northeast and West are least susceptible to sea-level rise impacts because of steeper average coastal profiles, geologic substrates composed of less erodible rock or glacial and riverine till, and lower rates of natural land subsidence. Coastal barrier islands and spits in the Northeast and low-lying salt marshes are exceptions in these regions; these areas are especially susceptible to erosion from storm surges associated with accelerated sea-level rise.

The most susceptible regions in the United States include the Gulf Coast, because of its relatively low-lying coastal topography and high existing rate of land subsidence, and the mid- Atlantic and south Atlantic areas, where low-lying coastal topography allows marine influence and hence sealevel rise to penetrate large distances inland. Extensive coastal lowlands that would be affected by sealevel rise are found in Louisiana and south Florida as well as eastern Texas, North Carolina, and the Chesapeake Bay of Maryland and Virginia. These coastal areas are fragmented by human use, such as urban settlements, resort towns, agriculture, and national seashores, which interact with physical effects to lead to a range of impacts.

Development, Demographic and Future Storm Damages

At the national level, more than half of the U.S. population currently lives in counties located along the 20,000 km of coastline. Projections of growth of the coastal population suggest that by 2010 the coastal population will have grown by 60 percent from 1960 levels. Florida is experiencing unprecedented population shifts as baby boomers enter retirement age and depart northern population centres for the southwest coast and the Miami-Ft. Lauderdale metropolitan region. Similarly, coastal resort communities such as Hilton Head and Myrtle Beach, South Carolina; the Outer Banks of North Carolina; and various communities in Georgia and along the Gulf of Mexico in Mississippi and Alabama are experiencing dramatic population growth. From 1950 to 1985, the coastal population of Texas increased 250 percent; Southern Californias population is expected to increase by 5.6 million people over the next 20 years.

Not surprisingly, the increase in coastal population has spurred a concomitant increase in population density, infrastructure, and property values that also contribute to the vulnerability to sealevel rise. Each week, about 8,700 new single-family homes are constructed along the U.S. coast. Moreover, use of coastal public lands and recreational resources has risen in step with population growth. In the decade from 1979 to 1989, recreational visits to coastal national parks, seashores, and monuments increased at a faster rate than coastal population itself.

Rates of population and property value growth in some coastal regions exceeded these national trends. Three of the important historical trends affecting the vulnerability of U.S. coastal regions to sealevel rise:

(1) size of coastal populations,

(2) value of insured coastal property, and

(3) amount of coastal wetlands.

Portions of the Gulf and Atlantic coasts, for example, have experienced the greatest proportional growth in numbers of people living close to the shoreline. Florida's population alone nearly tripled from 1960 to 1995, with much of that growth occurring since 1980. From 1988 to 1993, the total value of insured property in coastal counties from Maine to Texas increased 69 percent, from $1.9 trillion to $3.15 trillion, while the value of all insured U.S. properties, coastal and otherwise, increased about 65 percent.

Assessment of sealevel rise impacts should include assessment of vulnerability to storm damage. Coastal areas might be affected by climate change both through increased vulnerability to flooding and through effects on the intensity or frequency of storms. Recent tropical storms and hurricanes provide powerful evidence of the vulnerability of people and properties in U.S. coastal areas. Total damages from Hurricane Andrew in 1992, for instance, equalled about $30 billion, making it the most costly hurricane in U.S. history.

One reason for increased vulnerability to storms may be the interaction of the natural variability in storm intensity and its link to trends in coastal development. In the two or three decades prior to 1990, the eastern United States experienced a period of relatively mild hurricane activity. Perceptions of the risk of storms to property during this period may therefore have been underestimated. In the 1990s, as hurricane frequency and intensity increased to the higher end of the normal range, 20 to 30 years of relatively aggressive coastal development had left coastal regions much more vulnerable to storm damage. Analysis of landfalling hurricanes since 1925 indicates that seven hurricane seasons similar to the seasons experienced between 1940 and 1969 would have resulted in damages of $10 billion or more if they had occurred with 1995 patterns of coastal development. Future hurricane damages, projected from past storms, could average $5 billion per year. If climate change results in more frequent or more powerful storm events, damages could conceivably be even higher.

Ecological Services and Innovations

The myriad of cultural and aesthetic amenities as well as numerous valuable ecological services provided by coastal are as, al though not traded in economic markets, al so influence vulnerability. Seventeen million hectares of coastal marshes and wetlands in the United States remain today. Key ecosystem functions of these wetlands include: providing vital habitat and nursery grounds for various species of fish, shrimp, birds, and fur-bearing mammals; protecting uplands from saltwater intrusion and storm surges; and improving water quality through natural filtration of nutrients and toxic substances.

Coastal zones are also among the most biologically diverse areas in the United States. An evaluation of the viability of species in the coastal fringe by The Nature Conservancy shows the negative impacts of human development, pollution, and habitat fragmentation on coastal ecosystems. The Nature Conservancy found that 80 species and subspecies that exist only below the ten-foot contour of the U.S. coast are considered rare, imperilled, or critically imperilled.

The Science of Sea-Level Rise Assessment

Climate change could trigger a global rise in sea level by increasing the volume of water contained in the oceans' basins through thermal expansion of ocean water and the melting of polar ice and mountain glaciers. Thermal expansion would occur as higher global atmospheric temperatures over the next century warm the world's oceans, causing ocean water to expand. In addition, although the oceans contain most of the world's water, if all the ice on the earth's surface were to melt, global sea level would increase by about 100 meters. While 90 percent of the earth's frozen water is stored in the comparatively stable Antarctic ice sheet, other ice deposits are m o re susceptible to melting as a result of global warming.

Assessments of future changes in sea level require construction and implementation of a complex modelling framework, projections of future scenarios of major factors affecting climate change, and a clear characterisation of the uncertainty these analyses introduce. Using such a modelling framework, the Intergovernmental Panel on Climate Change concludes that increases in global temperatures over the next century could accelerate this rate of sealevel rise to an average of 5 millimetres per year, with a range of uncertainty of 2 to 9 mm/yr. The contribution of melting of Greenland ice to global sea levels is projected to be relatively small, while the Antarctic ice sheet is projected to increase in size with global warming because snowfall there will increase more than the ice will melt, lessening the overall sealevel rise.

Sea-Level Rise in the United States

While mean sea level rises and falls from year to year, and even from decade to decade, there is a clear long-term rising trend along most of the U.S. coast. IPCC concludes that there has been a global mean rise in sea level of between 10 and 25 cm over the last 100 years, representing the combined effect of an increase in ocean volume due to thermal expansion and the observed retreat of small ice caps and glaciers. Tide gauges have recorded relative sea levels in the United States for much of the 20th century. In a few locations, such as San Francisco, Key West, and New York, the data go back well into the 19th century. Data for New York City are shown in Figure 1.

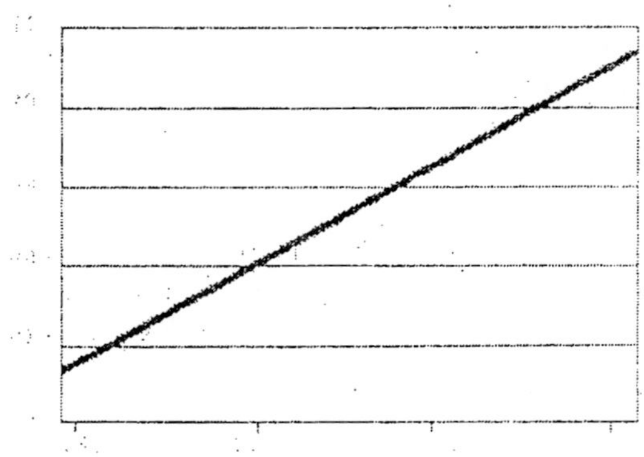

Figure 1 Measured Relative Sea Level at New York City

The combined effect of the global trend in sealevel rise and the contribution of geological processes that affect land elevation is illustrated. Relative sea level has been rising everywhere along the East and Gulf coasts at a rate of about 0.2 meter per century. The most rapid rise is in the Mississippi delta in Louisiana and the Chenier plain in east Texas, where relative sea levels are rising at up to 1 meter per century. The Mississippi delta and Chenier plain are composed of geologically young sediments that are still consolidating. This consolidation produces natural land subsidence rates that largely explain the observed high rates of relative sealevel rise.

The mid-Atlantic region is also seeing relatively rapid rises in sea level with rates of 0.3 to 0.4 meter per century at several stations. From Wilmington, North Carolina, to southern New Jersey, geophysical models suggest that postglacial rebound, which causes land that was not formerly glaciated to sink once glaciers retreat, is contributing 0.1 to 0.2 meter per century to relative sealevel rise, approximately doubling the effect of global sealevel rise. On the West Coast, data are more scarce, but available evidence indicates both rising and falling trends. Specifically, the West Coast is on a geological plate boundary, and in some areas there is a tendency for uplift to occur as one plate pushes up another, counteracting global sealevel rise. La Jolla, San Diego, and Seattle show relative sealevel rises of about 0.2 meter per century, similar to the East Coast. From northern California to Washington, relative sea level is falling at 0.05 to 0.16 meter per century.

Regional or local land subsidence also occurs. In some areas, groundwater or oil withdrawal has enhanced subsidence. Around Houston, Texas, 13,500 km^2 has subsided more than 30 cm in the 20th century. Around Galveston Bay, this subsidence necessitated coastal abandonment or increased coastal protection. In Long Beach, California, oil extraction produced land subsidence up to 9 meters.

Global Sea-Level Rise over the Next Century

Climate—induced increases in sea level from 1990 are projected to be 23 cm by 2050 and 55 cm by 2100. More recent work using new greenhouse gas emissions scenarios to model the effect of a warmer climate on global mean sea level shows a slightly higher rate of sea-level rise. IPCC estimates of global rises in sea level suggest a broad continuation of present trends up to a four- to tenfold acceleration in rates of global sea-level rise by the end of the century, with a mid-range estimate of two- to fivefold acceleration. Titus and Narayanan, on the other hand, conducted a probability-based analysis of global sea-level rise, based on a survey of the opinions of climate modellers. Their median estimate is that global sea levels would rise 34 cm from 1990 to 2100, with a 99 percent confidence interval falling between a decrease of 1 cm and a rise of 104 cm.

Gregory and IPCC noted a further complication: on a regional scale there will be departures from the global rise due to factors such as changes in ocean circulation, wind and pressure patterns, and ocean-water density. These factors will produce a regionally variable rise in the ocean surface around the global mean rise already discussed, and are additional factors to consider when constructing sealevel rise scenarios for impact assessment. However, this effect is still being quantified in modelling experiments and there are no widely-accepted scenarios available at present for impact analysis or coastal planning purposes.

Low, medium, and high relative sealevel rise scenarios over a long-term planning horizon the East and Gulf coasts, combining the global sealevel rise estimates from IPCC with expected land elevation changes. These estimates correspond to a continuation of present historical trends, a 20-cm global rise, and a 40-cm global rise. The "hot-spots" of coastal Louisiana/east Texas and the mid-Atlantic region are apparent, with relative sealevel rises up to 100 cm in the first region and 60 cm in the latter region. Elsewhere, the rise could be up to 50 cm, with a higher uncertainty in the area between Galveston and Port Isabel, Texas, owing to uncertainties in the rate of subsidence. Note that even the low-rise scenario produces significant rises in sea level.

These data indicate that large uncertainties exist about future global and regional sea levels. This uncertainty is unlikely to be substantially reduced in the near future. When estimating possible impacts, uncertainty can be handled by scenario-based vulnerability assessment. The range of estimates that emerges from this approach reflects the outcome of alternative model inputs. An alternative approach is to consider the probabilistic long-term sealevel rise scenarios for coastal planning, for example, those provided by Titus and Narayanan. However, assigning probabilities to alternative outcomes given the current state of knowledge is subjective. Considering all the uncertainties outlined above, different researchers could estimate significantly different probabilities.

Climate Change on Coastal Resources

Impacts of Relative Sea-Level Rise

The physical impacts of relative sea -level rise will vary by location and depend on a range of physical and socioeconomic factors, including human response. There are already widespread problems on the East and Gulf coasts linked by varying degrees to observed sea-level rise in the 20th century and often exacerbated by poor management. Most beaches are eroding and extensive tidal wetland losses are occurring, particularly in Louisiana and around the Chesapeake Bay. Recent hurricanes such as Hugo and Andrew have caused multi-billion-dollar losses, and freshwater supplies in the lower reaches of rivers can be adversely affected during periods of low flow.

In the Mississippi delta, roughly 40 km^2 of wetlands are lost every year. While relative sea-level rise is a major cause of these changes, attempts to maintain navigation channels and other human management of the delta have removed the sediment supplies that help the wetlands maintain elevation and keep pace with sea-level rise, as well as severely altering their hydrology. On the open coast, poor management of beach sand and its movement along shorelines has contributed to substantial erosion. Beach erosion will increase and the tourist industry will face the dilemma of maintaining beaches through the costly process of replacing sand or staging a planned re treat. The availability of sufficient sand and the question of who pays for beach nourishment will be key issues to be assessed. Accelerating sea-level rise would intensify all of these problems.

National assessments suggest that a one-meter rise in global sea levels could have significant impacts, including the inundation of about 35,000 km^2 of land divided almost equally between wetlands and upland. The 100-year coastal flood plain also could increase by 38 percent, or at least 18,000 km^2. Estimates of land inundated by a 0.5-meter global sealevel rise are about 24,000 km^2, which is still two thirds the inundation estimated under a one-meter rise because of the distribution of coastal elevation.

Major cities such as New Orleans, Tampa, Miami, Baltimore, Philadelphia, New York, Boston and Washington, DC, will have to upgrade flood defenses and drainage systems or face adverse consequences. New Orleans is particularly threatened by the loss of the surrounding wetlands in the Mississippi delta, which currently reduce the flood risk to the city from hurricane storm surges.

The extent and geographic distribution of wetlands losses depends on physical and human factors as well. The location of coastal wetlands is linked to present sea level; changes in sea level can cause wetlands to migrate landward. Wetlands with limited sediment supplies and low tidal range appear to be most threatened by sealevel rise. The availability of low-lying upland areas landward of the wetlands is also critical for wetland survival. If the uplands are protected by coastal defenses, landward migration is not possible. Titus et al. estimate that a one-meter rise in sea level would cause the loss of about 16,000 km^2 of wetlands, and if no wetland migration were possible, this loss would rise to 22,000 km^2. Loss estimates under a 0.5-meter rise are about two-thirds those under the one-meter scenario. Global analyses identify the Atlantic coast of North America as one of the more sensitive regions to wetland losses from sealevel rise. Where wetlands support commercial fisheries, that resource could be dramatically affected.

Catastrophic Melting Effects

If the West Antarctic ice shelf were destabilised by global warming and slid into the ocean, there would be a 5 to 7-meter rise in global sea levels. There is geological evidence that this scenario occurred about 100,000 years ago. While IPCC reviewed the available evidence and concluded that such an event is unlikely in the 21st century, the impacts of catastrophic melting of ice sheets could be enormous. Large parts of the East and Gulf coasts, particularly in Louisiana and Florida, are beneath a 5 to 7-meter elevation. While the risk of this type of event is extremely low it would be prudent to continue research to better understand the likelihood of such changes, and consider the needs of appropriate monitoring systems for Antarctica

Storm Intensity, Frequency, and Track Changes

The number, track, rainfall quantity, and intensity of storms and hurricanes might al so change with global warming, although future patterns of storms and hurricanes are uncertain Generally, climate-forecasting tools used in climate studies do not have fine enough spatial resolution to be able to simulate individual tropical storms. There is some empirical evidence that the frequency of Atlantic hurricanes might increase as sea surface temperatures increase because of the positive correlation between these temperatures and hurricane frequency.

Based on present understanding, IPCC has concluded that future patterns of storm frequency, track, and intensity are uncertain. Both increases and decreases in storminess are possible, and changes are likely to differ among regions around the world. Given present variability in the occurrence and strength of storms, it might be difficult to observe long-term trends. Based on an analysis summarised in Pielke and Landsea, 11 major hurricanes hit the East Coast, including the Florida Peninsula, from 1941 to 1965. From 1965 to 1990, when the populations of Florida and other southern states grew enormously, only two major hurricanes struck the East Coast and none struck Florida.

In the 1990s, hurricane frequency returned to more typical levels, and the risk to coastal property is more present in the public mind. Future hurricane damages, if projected based on past storms, could average $5 billion per year. If climate change results in more frequent, more powerful, or wetter storm events, or, if it substantially changes the track of future hurricanes toward more northerly locations, damages could increase. Even if climate change has no major long-term effects on the occurrence of storm effects, the variability in storm and hurricane occurrence discussed above will interact with sealevel rise. Therefore, understanding and predicting this variability is also an important topic for further research.

Adaptation to Coastal Threats and Human Response

Three options are available to decision-makers who contemplate responding to coastal threats, each implying trade-offs in the distribution of costs and benefits: protection, accommodation, and planned retreat. Protection seeks to exclude the hazard, accommodation allows human activities and the hazard to coexist, while planned retreat removes human activity from the hazardous zone. Protection options include hard structure responses and "soft" engineering responses that utilise sediment.

These strategies could protect some of the resources vulnerable to sea-level rise, but might sacrifice other resources in the process. Building hard structures, for example, would limit the ability of beach and wetland resources to migrate inland as sea-level rises. These losses are often termed "coastal squeeze".

Accommodation might include elevation of buildings, modification of drainage systems, or land-use changes. Planned retreat could involve such strategies as coastal development restrictions. In both cases, coastal squeeze is avoided.

In the United States, most decisions that involve a structural response are made at the local level, sometimes on the basis of cost-benefit analysis. State and national institutions can also play a role, particularly by providing appropriate guidance and incentives. In all cases, the distribution of costs and benefits is a critical factor. Costs of protection are typically borne collectively, while potential damages threaten individuals or their immediate communities.

The interests of individuals with property at risk can play a disproportionate role in the framing of policy and its implementation. Overall, the particular strategy adopted in response to perceived threats from sealevel rise depends on many factors. These factors include the value of the land or infrastructure under threat, the financial and economic re sources that can be brought to bear, the local landscape of coastal management policy, and the ability to understand and implement adaptation options.

Protection of Hard Structure

Seven published studies have offered specific cost estimates for various protective structures. Estimates of their fixed construction costs for dikes or levees built to protect against a onemeter rise in sea level range from $150 to $800 per linear foot. Corresponding cost estimates for sea wall and bulkhead construction range from $150 to $4,000 per linear foot.

The range in costs reflects location-specific factors such as the amount of site and foundation preparation work necessary, drainage requirements, and differences in materials and labour costs. Much of the sea-level rise impacts literature uses the median estimate of $750 per linear foot drawn from Gleick and Maurer.

Hard structures do not have to be constructed until they are needed. Ideally, hard structures would be built just in advance of the threat of inundation. Because the pace of greenhouse induced sealevel rise is unknown, some anticipation and monitoring is required to implement appropriate pre-emptive responses.

Yohe and Neumann assessed the timing decision and developed the notion of a warning threshold that relates anticipated sealevel rise through 2100 to the year of potential need for hard structure protection. Given the local subsidence in this area, the threat of inundation appears when relative sealevel rises 62 cm. Combining this warning threshold with linear sealevel rise trajectories produced results for a range of high sealevel rise scenarios. In this example, 2017 is the earliest date of concern, and then only if sealevel rise of 110 cm is expected through 2100.

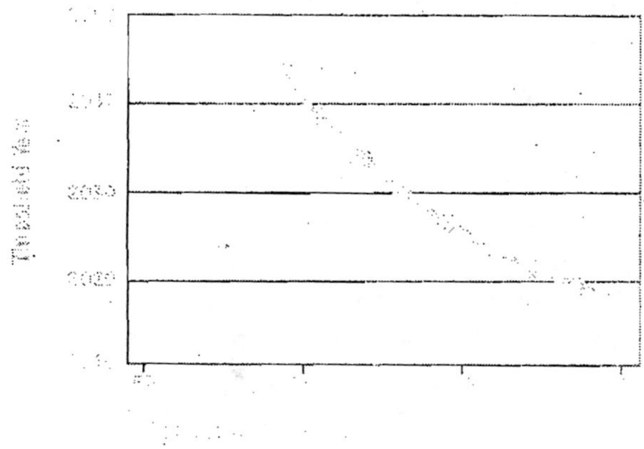

Figure 2 Results of Warning Threshold Calculation for Charleston, South Carolina

Improved Sediment Management

Since hard protection in response to an erosion problem does not solve the fundamental problem of adiminishing sediment resource, an alternative approach is artificial addition of sediment and /or improved sediment management, or soft protection, to increase the size of the beach or wet land. The added sediment is generally sacrificed as the erosive processes continue; therefore repeated additions of sediment are required.

Beach nourishment is a popular and cost effective option in highly developed areas with popular beaches and valuable beachfront real estate, especially during the early onset of erosion. Florida and New Jersey, for example, have established funds for a variety of beach nourishment projects. However, the long-term effectiveness of beach nourishment remains uncertain due to an incomplete understanding of coastal processes and how they will be influenced by sealevel rise and climate change.

Adequate monitoring of beach nourishment projects is essential to understanding technical issues concerning their longevity and effectiveness.

While many experts agree that beach nourishment with regular re-nourishment cycles is an appropriate response to sealevel rise, it remains unclear when nourishment would cease to be cost-effective and another response strategy would be required. This strategy is a site-specific function of the rate of sealevel rise, the unit cost of sand, and the length of beach being nourished. As sea levels rise more rapidly, and the cheaper sand resources are exhausted, nourishment is likely to cease to be cost-effective in some coastal locations. A move to hard protection will eventually degrade the beach, as the sand is scoured away by wave action, and hence may imply loss of recreational and ecological resources.

As nourishment has become more popular, it has shown the value of coastal sediment. Therefore, in addition to nourishment, the importance of bypassing sand at inlets and beneficial uses of dredge spoil have received greater priority. Sealevel rise reinforces this trend and suggests that soft defenses should be placed in a broader context of sediment management. This approach would comprise a regional understanding of sediment fluxes, including human interventions. Efforts in other countries towards improving shore line management based on coastal cells may be instructive. Soft protection strategies there fore raise a number of important trade-offs which require urgent attention.

Primary Tools for Accommodation and Planned Retreat

The primary tools for accommodation and planned retreat are land use and development planning. Setback measures, a zoning mechanism for planned retreat employed by states, require that new structures be set back from the shore, usually by some multiple of the average annual rate of erosion. Nine states require new construction to be set back by a distance equivalent to the extent of erosion expected over the next 30 to 60 years. Maine's Dune Rules reference setbacks explicitly to sea-level rise by requiring demolition of large structures if the sea level rises by one meter or more. These types of measures can be controversial and legally contentious, but once implemented they are an effective means of communicating and enforcing a planned retreat strategy. Post-disaster reconstruction plans are a second type of mechanism for planned retreat.

Post-disaster plans limit or prohibit reconstruction of coastal property severely damaged by hurricanes, storms, or other episodic flooding. An example is the federal National Flood Insurance Programme (NFIP), which provides subsidized insurance for damage to coastal structures due to flooding or coastal erosion. NFIP requires elevation of structures damaged more than 50 percent of pre-storm value above the 100-year flood level plus wave heights. In practice, in those states with coastal management policies that feature setback rules, there has been considerable public resistance to their implementation even for existing erosion problems. Planning for sealevel rise implies larger setbacks might be required. In other instances, case law may dictate that setback rules re p resent an unconstitutional taking of property. Both of these issues may limit the effectiveness of these planned re treat policies.

As evidence of the difficulty of implementing setbacks, owners of beachfront property in South Carolina were able to either circumvent the setbacks or showed a reluctance to consider the risks of sealevel rise when making decisions about whether to rebuild after Hurricane Hugo in 1989. Most owners repaired or rebuilt their homes despite extensive damage to their structures, and presumably with the full knowledge of the risks of proximity to the shoreline.

At the federal level, the Coastal Zone Management Act (CZMA) encourages better management and planning for nonfederal coastal resources by state and federal agencies. In response to the CZMA, most states have developed their own coastal zone management programmes. In some cases, states have enacted legislation to explicitly address the potential for accelerated sealevel rise. The South Carolina Beachfront Management Act, for example, includes provisions for identifying critical erosion areas and for regulating construction along the coastline.

One method of accommodating sealevel rise is to elevate structures. This option has been implemented by individual property owners in places such as the New Jersey and South Carolina shorefronts, and provides an effective means of responding to concerns about increased frequency of episodic flooding associated with sealevel rise. This strategy does nothing to preserve the beach, however. Raising land elevations can preserve beaches, where the cost of sediment placement is not prohibitive, and has been considered for highly valued coastal areas such as New Jersey's Long Beach Island.

Economic Impacts on Coastal Properties and Wetlands

IPCC suggests impact assessments should reflect at least five categories of impacts: inundation and erosion of property, inundation of wetlands, effects on recreation, effects on drinking water quality and quantity, and effects on port infrastructure. The potential costs associated with all of these impacts are interrelated, although they are often assessed separately. Property losses dominate cost estimates in the United States, but the wetland acreage lost is comparable to upland acreage lost. The monetary value of wetland losses, currently not included in most impact assessments, may prove to be large. As are sult, major estimation difficulties must be overcome. Likewise, national assessments of the effects of sea-level rise on recreation, water, and ports have not been conducted. Some portion of these effects may be captured, however, in more careful studies using property value changes to measure economic impact.

Economic Cost to Developed Property

The magnitude of developed property impact estimates reflects two key factors: the sea -level rise scenario and the modelling of responses. Some of the earliest results, such as those produced by Schneider and Chen, reflected extreme sea-level rise scenarios, 450 to 750 cm by 2100, that are irrelevant by current standards. The first systematic national study of more moderate sea-level rise scenarios was the U.S. Environmental Protection Agency's 1989 Report to Congress. For the EPA study, Yohe estimated the costs of allowing sea level to inundate low-lying property along the much more reasonable global sealevel rise scenarios of 50 cm and 100 cm through 2100. Yohe's results indicate the value of property that could be inundated at $138 billion for a 50-cm rise and $321 billion for a 100-cm rise.

As part of the same EPA study, Titus estimated the cumulative cost of protecting all developed property along the U.S. coastline resulting from a 100-cm rise at $73 billion to $111 billion. Nordhaus used the EPA report to estimate an annual cost of protection of $4.9 billion and an annual cost of lost land of $2.4 billion, consistent with a scenario of seas rising roughly one meter by 2100. However, both the Yohe and Titus estimates reflected simple decision rules for the response—either protect or abandon all developed property. Neither estimate reflected the fact that for highly valued areas protection is the least cost response, while for lesser valued areas abandonment is less costly.

Later studies suggest that taking adaptation into account could matter significantly. Specifically, in a global study, Fankhauser addressed a key issue about whether protecting individual segments of coastal property reflected local conditions and consideration of the relative costs and benefits of protection. Fankhauser assumed gradual inundation patterns and reported amortised and cumulative protection costs of $570 million for a 50-cm rise and $1 billion for a 100-cm rise by 2100. Thus including adaptation for a 100-cm rise resulted in an estimate one-fifth that of Nordhaus' estimate, which did not include adaptation.

Yohe et al. followed with a micro-level assessment of a sample of U.S. coastal sites that reflected the variability in local topography and baseline land use. This assessment modelled adaptation decisions to protect or abandon property assuming a cost-benefit decision framework at a relatively high resolution within a sample of U.S. coastline. The sums of protection costs and the contemporaneous values of lost property were reported for each year though 2100.

The annual estimates for this study are between one-tenth and one-fifth of the Fankhauser estimates, although property losses were added to protection costs. These lower costs result from incorporating a local decision-making framework for determining the response. Specifically, at the scale of the Yohe et al. study, some property is not vulnerable to rising seas because of abrupt cont-ours, while other property is not very valuable because it is already susceptible to tidal flooding. By accommodating both of these effects, the cost estimates of Yohe et al. were much lower than Fankhauser's even though they were based on property values designed to change over time in line with anticipated increases in future per capita income and population growth.

Yohe and Schlesinger used the Yohe et. al. methodology to produce a representation of transient protection costs and property losses as a function of sealevel rise and greenhouse gas emissions trajectories. Their work incorporated three different assumptions about future sulfate emissions, seven probabilistically weighted carbon and associated gas emissions trajectories, three different climate sensitivities to greenhouse gas concentrations, and three different forcing coefficients for sulphates to produce a probabilistically weighted range of "not-implausible" sealevel rise trajectories.

The Yohe and Schlesinger results are 30 percent higher than comparable expected values computed under the assumption of perfect foresight—a difference that Nordhaus has recently offered as a first representation of the extra cost of climate surprise. Nordhaus' point is that adjustment to moderate sealevel rise will not be terribly expensive to market economies if the change is foreseen. The Yohe and Schlesinger results also illustrate the range of economic impact results that are possible using alternative sealevel rise scenarios. The results indicate that economic impacts to developed property are not proportional to anticipated sealevel rise. Specifically, the most moderate scenarios yield much smaller than expected results, implying that the impacts of sealevel rise at the low end of current estimates could be as small as $10 million per year in 2065.

As illustrated by recent work, Yohe's series of estimates relying on a cost-benefit decision model provide valuable insights for policy-makers. However, these studies have been criticised for being too optimistic about the potential for adaptation and optimal decision-making at the local level. There are two main classes of criticism. First, long-term costs and benefits over the life of a project are difficult to estimate at the local level, so the cost-benefit framework may not be a practical model of local decision-making. Second, the cost-benefit decision model used does not incorporate the influence of storms on development patterns over time. In a local case study, West and Dowlatabadi modelled the effects of storms and erosion using a model of random storm intensity.

The West and Dowlatabadi model incorporates market valuation and private investor decisions into an analysis of a hypothetical coastal community with and without sealevel rise. They begin with the methodology of Yohe et al. so that the incremental effect of storms and erosion can be measured directly. West and Dowlatabadi conclude that expected costs attributable to sealevel rise are small. They also observe, though, that actual damages could become more significant under different assumptions about the geographical distribution of property values, accelerated dune erosion, or even unlucky sequencing of coastal storms, increasing total costs attributable to sealevel rise by almost 20 percent.

The overall conclusion from this body of work is that as estimates of sealevel rise have moderated, and as the ability to model adaptation and incorporate local conditions has increased, estimates of the impact of sealevel rise on coastal property have moderated. However, attaining economically efficient adaptation will require a widespread understanding of sealevel rise that does not yet exist. While it is clear that a significant magnitude of investment in coastal structures is at risk from sealevel rise, uncertainty as to the ultimate response affects the magnitude of impacts. Estimates a 50 cm sealevel rise by 2100 could cause cumulative impacts to coastal property in the United States of $20 billion to roughly $150 billion. The large difference between estimates that reflect no adaptation, and the more recent studies that reflect economically efficient adaptation suggest that there should be major efforts to encourage understanding of strategies to more efficiently respond to sealevel rise.

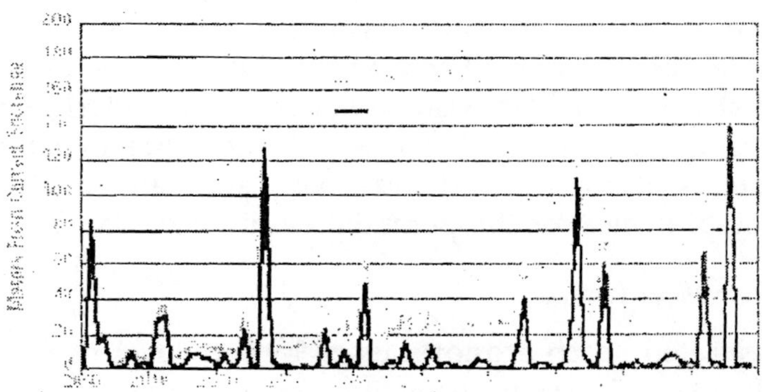

Figure 3 Simulated Inundation From Coastal Storms with and without Sea-Level Rise

Assessment of Wetlands

Existing assessments of lost wetland are as, while only first approximations, nonetheless point to the potentially large magnitude of damages that sea -level rise could cause to coast al wet l and resources Current estimates of the impact of wetland losses suffer from two critical methodological and data gaps. First, it is difficult to model the dynamic processes of wetland accretion and migration. As evidenced by historical data in south San Francisco Bay, accretion and migration could mitigate wetland losses in some cases. If these processes are constrained in these areas, however, wetland losses might increase due to coastal squeeze. Moreover, as suggested by historical rates of wetland loss in some areas, the beneficial accretion of sediment and biogenic material within wetlands may or may not keep pace with the rate of relative sealevel rise. If wetlands fail to keep pace with sea level by accretion and/or migration, they will be progressively degraded and ultimately destroyed and would no longer provide most of the services they currently provide.

Second, there have been many efforts to assign a dollar value to wetland resources over the last two decades, but the overall economic value of wetlands is not well-characterised by existing value estimation techniques. Currently, estimates of the total unit value of specific functions at specific wetlands exist, and can be used to develop a sense of the potential value of wetlands.

Recreation Losses

Accelerated beach erosion caused by sea -level rise might lead to large losses in recreation al value. Beaches may be able to migrate inland, mitigating losses, but may still lose value. Another scenario would be the construction of hard structures to protect beachfront property, which might lead to the progressive loss of the beach. Some of these losses could be mitigated through beach nourishment. While there have already been numerous nourishment projects around the U.S. coast, the scale of nourishment required to hold the line for the next century could be vast, even if sea-level rise does not accelerate.

In all instances, the costs of adapting to rising sea levels, which may be quite high in certain cases, must be considered. Therefore, while certain beaches, such as Miami Beach, are likely to be nourished irrespective of the sea-level rise scenario, less densely developed coastal areas may have to consider a managed re treat in the face of sea-level rise. Strategic assessments of potential demands for nourishment over the next century given a range of sea-level rise scenarios would be helpful. These assessments might be best accomplished at the state level.

At least a portion of the recreational impact should be reflected in the estimates of lost property value since the value of coastal recreation should be closely linked to the value of nearby structures. In other words, the value of coastal property includes a substantial premium that is associated with access to the coast's recreational amenities. An issue for future research is whether the methods for estimating losses in property value associated with sealevel rise also capture any reduction in the recreational amenity.

Challenges of Future Assessments

Future assessments are faced with the daunting challenge of acknowledging the complexities of sea -level rise impacts as they attempt to characterise climate change impacts and recommend policy responses appropriate for the U.S. coast. The context of the impacts of sea-level rise is complex. The risks posed by sea-level rise to the coastal zone may sometimes be considered more straight forward to assess than other climate change impacts. Unlike these other impacts, sea-level rise occurs in one dimension with some confidence about the direction of change, if not always the magnitude.

Yet, sea-level rise occurs within the already dynamic geology of the coastal zone. Responding to any rise will require balancing multiple and sometimes competing uses and values. It also will involve planning construction and protection of long-lived capital assets. Responding to sea-level rise in the most economically efficient manner will challenge humans to learn and adapt over a long period of time.

There has been some progress in defining coastal management policies that can enhance the ability of coasts to adapt to sealevel rise. Nicholls and Branson describe research efforts focusing largely on long-term coastal planning not only for uncertain sealevel rise but also for weather events that can cause enormous damage. They introduce the concept of natural and human resilience to routine and extreme events. Klein et al. cast the issue clearly in their review of policies of the Netherlands, which has a law that forbids coastal erosion—an extreme manifestation of humans standing in the way of natural forces.

Klein et al. criticise this static approach and advocate more flexibility, recognising that coping with coastal dynamics could provide more safety at a lower cost. They argue that systems designed to handle only the gradual consequences of sealevel rise may be more vulnerable to the disastrous consequences of extreme events like storms. Enhancing coastal resilience can be an effective response to an uncertain future, including sealevel rise.

7

Effects of Climate Extremes

A range of physical, ecological and social mechanisms can explain an association between extremes of climate and disease. Social mechanisms may be very important but are difficult to quantify: for example, droughts and floods often cause population displacement. Outbreaks of infectious disease are common in refugee populations due to inadequate public health infrastructure, poor water and sanitation, overcrowding and lack of shelter. Climate also can affect infectious diseases that are spread via contaminated water or food. Water-related diseases are a particular problem in poor countries and communities, where water supplies and sanitation often are inadequate. Outbreaks of cholera, typhoid and diarrhoeal diseases can occur after flooding if the floodwaters become contaminated with human or animal waste, while drought reduces the water available for washing and sanitation and also tends to increase the risk of disease.

El Niño and Diseases

There is a well-studied relationship between rainfall and diseases spread by insect vectors which breed in water, and are therefore dependent on surface water availability. The main species of interest are mosquitoes, which spread malaria and viral diseases such as dengue and yellow fever. There is considerable evidence linking mosquito abundance to rainfall events. Mosquitoes need access to stagnant water in order to breed-conditions that may be favoured by both wet and dry conditions. For example, heavy rain can create as well as wash away breeding sites, while in normally wet regions drought conditions can increase breeding sites by causing stagnation of water in rivers.

The timing of rainfall in the year and the co-variation of other climate factors also are likely to be important. Vector-borne disease transmission is sensitive to temperature fluctuations also. Increases in temperature reduce the time taken for vector populations to breed. Increases in temperature also decrease the incubation period of the pathogen meaning that vectors become infectious more quickly.

Effect of Malaria

Malaria is the world's most important vector-borne disease. Over 2.5 billion people are at risk, and there are estimated to be 0.5 billion cases and more than 1 million deaths from malaria per year. Malaria incidence is influenced by the effectiveness of public health infrastructure, insecticide and drug resistance, human population growth, immunity, travel, land-use change and climate factors.

Very high temperatures are lethal to the mosquito and the parasite. In areas where temperatures are close to the physiological tolerance limit of the parasite, a small temperature increase would be lethal to the parasite and malaria transmission would therefore decrease. However, at low temperatures a small increase in temperature can greatly increase the risk of malaria transmission. Malaria's sensitivity to climate is illustrated in desert and highland fringe areas where rainfall and temperature, respectively, are critical parameters for disease transmission. In these regions higher temperatures and/or rainfall associated with El Niño may increase transmission of malaria. In areas of unstable malaria in developing countries, populations lack protective immunity and are prone to epidemics when weather conditions facilitate transmission.

Drought in the previous year has been identified as a factor contributing to increased malaria mortality. There are several possible reasons for this relationship. Drought-related malnutrition may increase an individual's susceptibility to infection. Also, drought may reduce malaria transmission resulting in a reduction in herd immunity in the human population. Therefore, in the subsequent year the size of the vulnerable population is increased. Alternatively, a change in ecology of the natural predators may affect mosquito vector dynamics; mosquito populations recover more quickly than their predator populations following a dry year.

Famine conditions may have contributed to excess mortality during historical epidemics of malaria, for example following the 1877 El Niño in India. Many deaths occurred after the end of the drought; the proximate cause was malaria when drought-breaking rains increased vector abundance, exacerbated by population movement and the concentration of people in feeding camps.

Many parts of South America show ENSO-related climate anomalies. Serious epidemics in the northern countries of South America have occurred mainly in the year after El Niño (year +1). In 1983 following a strong El Niño event, Ecuador, Peru and Bolivia experienced malaria epidemics. In Venezuela and Colombia, malaria increased in the post-Niño year (+1). A statistically significant relationship was found between El Niño and malaria epidemics in Colombia, Guyana, Peru, and Venezuela. The causal mechanisms are not completely understood. El Niño is associated with a reduction of the normal high rainfall regime in much of Colombia, as well as an increase in mean temperature, increase in dew point, and decrease in river discharges. These relationships between malaria and ENSO nevertheless can be used to predict high and low-risk years for malaria, giving sufficient time to mobilise resources to reduce the impact of epidemics.

Other researchers emphasize the significance of non-climate factors in explaining recent malaria epidemiology in Africa. A resurgence of malaria in the highlands of Kenya over the past 20 years has been attributed to resistance to antimalarial drugs. Another study did not find a relationship between climate trends and the timing of malaria epidemics in Kenya. Based on a 30-year time series of climate and disease data, it concluded: "... intrinsic population dynamics offer the most parsimonious explanation for the observed interepidemic periods".

One study has reported no significant meteorological trends in four highaltitude sites in East Africa where increases in malaria have been reported. This study used spatially averaged climate data that may be unreliable for this purpose. An association between rainfall, temperatures and the number of inpatient malaria cases three to four months later has been reported recently.

Effect of Dengue

Dengue is the most important arboviral disease of humans, occurring in tropical and subtropical regions worldwide. In recent decades, dengue has become an increasing urban health problem in tropical countries. The disease is thought to have spread mainly as a result of ineffective vector and disease surveillance; inadequate public health infrastructure; population growth; unplanned and uncontrolled urbanisation; and increased travel. The main vector of dengue is the domesticated mosquito, Aedes aegypti, that breeds in urban environments in artificial containers that hold water. Dengue also can be transmitted by Aedes albopictus, which can tolerate colder temperatures.

Dengue is seasonal and usually associated with warmer, more humid weather. There is evidence that increased rainfall in many locations can affect the vector density and transmission potential. ENSO may act indirectly by causing changes in water storage practices brought about by disruption of regular supplies. Rainfall may affect the breeding of mosquitoes but this may be less important in urban areas: Aedes aegypti breed in small containers, such as plant pots, which often contain water in the absence of rain.

Between 1970 and 1995, the annual number of epidemics of dengue in the South Pacific was positively correlated with the Southern Oscillation Index (SOI). This is plausible since, in this part of the world, high positive values of the SOI are associated with much warmer and wetter conditions than the average-ideal for breeding of mosquitoes. In a subsequent study, Hales et al. examined the relationship between ENSO and monthly reports of dengue cases in 14 island nations in the Pacific. There were positive correlations between SOI and dengue in ten countries.

In five of these there were positive correlations between SOI and local temperature and/or rainfall. During La Niña, these five islands are likely to experience wetter and warmer than normal conditions. Local weather patterns may trigger an increase in transmission in larger, more populated islands where the disease is endemic, but infected people then carry the disease to smaller neighbouring islands.

Effect of Rodent-borne Diseases

Rodents act as reservoirs for a number of diseases whether as intermediate infected hosts or as hosts for arthropod vectors such as ticks. Certain rodentborne diseases are associated with flooding including leptospirosis, tularaemia and viral haemorrhagic diseases. Other diseases associated with rodents and ticks include plague, Lyme disease, tick borne encephalitis (TBE) and hantavirus pulmonary syndrome (HPS). Rodent populations have been shown to increase in temperate regions following mild wet winters. One study found that human plague cases in New Mexico occurred more frequently following winter-spring periods with above average precipitation. These conditions may increase food sources for rodents and promote breeding of flea populations. Ticks also are climate sensitive.

Infection by hantaviruses mainly occurs from inhalation of airborne particles from rodent excreta. The emergence of the disease hantavirus pulmonary syndrome in the early 1990s in the southern United States has been linked to changes in local rodent density. Drought conditions had reduced populations of the rodents' natural predators; subsequent high rainfall increased food availability in the form of insects and nuts. These combined effects lead to a tenfold increase in the population of deer mice from 1992 to 1993. In 1998, an increase in cases of hantavirus was linked to increased rodent populations which, in turn, were attributed to two wet, relatively warm winters in the southern United States associated with 1997/98 El Niño.

Effect of Diarrhoeal Diseases

Many enteric diseases show a seasonal pattern, suggesting sensitivity to climate. In the tropics diarrhoeal diseases typically peak during the rainy season. Floods and droughts are each associated with an increased risk of diarrhoeal diseases, although much of the evidence for this is anecdotal. The suggestion is plausible, however, since heavy rainfall can wash contaminants into water supplies, while drought conditions can reduce the availability of fresh water leading to an increase in hygiene-related diseases. Major causes of diarrhoea linked to contaminated water supplies are: cholera, cryptosporidium, E.coli, giardia, shigella, typhoid, and viruses such as hepatitis A. Outbreaks of cryptosporidiosis, giardia, leptospirosis and other infections have been shown to be associated with heavy rainfall events in countries with a regulated public water supply.

An association between drinking water turbidity and gastrointestinal illness has been reported. This was one of the first studies to apply time series methods to the analysis of water-related disease. A study of waterborne disease outbreaks in the United States has shown that about half were significantly associated with extreme rainfall. Outbreak locations from an Environmental Protection Agency database were assigned to watersheds. The rainfall in the month of the outbreak and in previous months was estimated from climate records: for outbreaks associated with surface water the association was strongest for rainfall events in the same month as the outbreak.

Transmission of enteric diseases may be increased by high temperatures, via a direct effect on the growth of disease organisms in the environment. In 1997 a markedly greater number of patients with diarrhoea and dehydration were admitted to a rehydration unit in Lima, Peru, when temperatures were higher than normal during an El Niño event. A time series analysis of daily data from the hospital confirmed an effect of temperature on diarrhoea admissions, with an estimated 8% increase in admissions per 1°C increase in temperature.

Temperature Related: Heatwaves and Cold Spells

In recent years there has been a great increase in interest in time series studies of temperature and mortality. These are seen as the most satisfactory method for quantifying the short-term associations between ambient temperatures and daily mortality. Any long-term patterns in the series are removed. The effect of a hot day is apparent only for a few days in the mortality series; in contrast, a cold day has an effect that lasts up to two weeks. In many temperate countries mortality rates in winter are 10-25% higher than death rates in summer but the causes of this winter excess are not well understood.

It is likely that different mechanisms are involved in heat and cold related mortality; cold related mortality in temperate countries is related in part to the occurrence of seasonal respiratory infections. High temperatures cause some well-described clinical syndromes such as heatstroke. Very few deaths are reported as attributed directly to heat. Exposure to high temperatures increases blood viscosity and it is plausible that heat stress may trigger a vascular event such as heart attack or stroke.

Impact of Heatwave Events on Mortality

During heatwaves, excess mortality is greatest in the elderly and those with preexisting illness. Much of this excess mortality is due to cardiovascular, cerebrovascular and respiratory disease. The mortality impact of a heatwave is uncertain in terms of the amount of life lost: a proportion of the deaths occur in susceptible persons who were likely to have died in the near future. Nevertheless, there is a high level of certainty that an increase in the frequency and intensity of heatwaves would increase the numbers of additional deaths due to hot weather.

There is no standard international definition of a heatwave. Operational definitions are needed for meteorological services. As meteorological agencies are becoming more commercialised they are keen to develop practical applications of their forecasts and tailor them to user needs. The Netherlands meteorological bureau uses the following definition to trigger advance warnings in the media and directly to health services: at least 5 days with maximum temperature above 25°C of which at least 3 days with maximum temperature above 30°C.

The evidence on which this is based is not clear. In the United States, the National Weather Service suggest that a heat advisory be issued when the daytime heat index reaches 40.6°C and a night time minimum temperature of 26.7°C persists for at least 48 hours. Local definitions are used: in Dallas the medical examiners office define a heatwave as three consecutive days of temperatures over 37.8°C.

It is surprisingly difficult to define a heatwave as responses to very high temperatures vary between populations and within the same population over time. A 1987 heatwave in Athens resulted in 926 deaths classified as heat-related, although the attributable excess mortality was estimated to be more than 2000. A subsequent heatwave in 1988 was associated with a much smaller excess mortality. This has been observed also in Chicago following the 1995 heatwave.

Few analyses have looked at the impacts of heatwaves in developing countries and the evidence is largely anecdotal. A heatwave in India in June 1998 was estimated to have caused 2600 deaths over 10 weeks of high temperatures. In Ores, the temperature rose to 49.5°C and was reported to have caused 1300 deaths. The high temperatures were exacerbated by recurrent power failures that affected cooling systems and hospital services in Delhi.

Vulnerability to Temperature Mortality

Indicators of vulnerability to heat and cold that have been investigated include:

— age and disease profile

— socioeconomic status

— housing conditions

— prevalence of air conditioning

— behaviour

These factors also have counterparts in individuals as risk factors for heat related mortality or morbidity, such as presence of air conditioning at time of death. Both individual and population level studies provide strong and consistent evidence that age is a risk factor for heat-related mortality. Studies vary on the age at which the vulnerability is increased. There are physiological reasons why the elderly are more vulnerable.

An important study was undertaken following the Chicago heatwave in 1995. Semenza et al. interviewed the relatives of those who died during the heatwave and controls who lived near the case, matched for age and neighbourhood. Individual risk factors for dying in the heatwave were identified: chronic illness; confined to bed; unable to care for themselves; isolated; without air conditioning. A comparison of mortality rates in three Illinois heatwaves by age group, sex and ethnic group found that women and white people were at more risk.

Winter Mortality Rates

In many temperate countries there is a clear seasonal variation in mortality, death rates during winter being 10-25% higher than those in summer. The major causes of winter death are cardiovascular, cerebrovascular, circulatory and respiratory diseases. Annual outbreaks of winter diseases such as influenza, which have a large effect on winter mortality rates, are not strongly associated with monthly winter temperatures. Social and behavioural adaptations to cold play an important role in preventing winter deaths in high latitude countries. Sensitivity to cold weather is greater in warmer regions. Mortality increases to a greater extent with a given fall in temperature in regions with warmer winters, in populations with less home heating and where people wear lighter clothes.

The elderly are particularly vulnerable to winter death, having a winter excess of around 30%. This vulnerability is not yet well understood but may arise through a combination of physiological susceptibility, behavioural factors and socioeconomic disadvantage. Excess winter mortality is an important problem in the United Kingdom where there has been much debate about the role of poor housing, fuel poverty and other socioeconomic issues for the elderly population. Several studies have linked routine mortality data at ward or enumeration district level with small-area indicators of housing and deprivation.

Climate Change on Temperature Related Mortality

Global climate change is likely to be accompanied by an increase in the frequency and intensity of heatwaves, as well as warmer summers and milder winters. Extreme summer heat's impact on human health may be exacerbated by increases in humidity. There has been significant warming in most regions in the last 25 years some of which the IPCC has attributed to human activities. However, it is not clear that the frequency of heatwaves has been increasing, although few studies have analysed daily temperature data to confirm this. There is much regional variation in the trends observed. Gaffen and Ross looked at data from 1961-1990 for 113 weather stations in the United States and found that the annual frequency of days exceeding a heat stress threshold increased at most stations.

Predictive modelling studies use climate scenarios to estimate future temperature related mortality. Those studies which use the empirical statistical model find that reductions in winter deaths are greater than increases in summer deaths in temperate countries. However, other methods indicate a more significant increase in summer deaths. Kalkstein and Green estimated future excess mortality under climate change in United States' cities.

Populations can be expected to adapt to changes in climate via a range of physiological, behavioural and technological changes. These will tend to reduce the impacts of future increases in heatwaves. The initial physiological acclimatisation to hot environments can occur over a few days but behavioural and technological changes, such as changes to the built environment, may take many years.

While it is well established that summer heatwaves are associated with short term increases in mortality, the extent of winter-associated mortality directly attributable to stressful weather is difficult to determine and currently being debated. Limited evidence indicates that, in at least some temperate countries, reduced winter deaths would outnumber increased summer deaths. The net impact on mortality rates will vary between populations.

Consequences of Natural Disasters

The health effects of disasters are difficult to quantify because secondary effects and delayed consequences are poorly reported and communicated. Information on natural disasters generally is gathered by the organisations and bodies directly involved in disaster relief and reconstruction. As a result, information usually is collected for specific operational purposes not as a database; figures are estimated, not measured directly.

This is especially true of flood events and windstorms where the actual deaths and injuries directly caused by the event are small compared to the problems that arise as a result, including deaths from communicable diseases and the economic losses sustained.

El Niño has an effect on the total number of persons affected by natural disasters. Worldwide, disasters triggered by droughts are twice as frequent during the year after the onset of El Niño than other years. This risk is concentrated in southern Africa and southeast Asia. The El Niño effect on disasters is strong enough to be apparent at the global level. In an average El Niño year, around 35 per 1000 persons are affected by a natural disaster.

Consequences of Weather Disasters

Developing countries are poorly equipped to deal with weather extremes. The number of people killed, injured or made homeless by natural disasters is increasing alarmingly. This is due partly to population growth and the concentration of population in high-risk areas like coastal zones and cities. Large shanty-towns with flimsy habitations often are located on land subject to frequent flooding. In many areas the only land available to poor communities may be that with few natural defences against weather extremes.

Direct hits of extreme events on towns and cities tend to cause large losses. In recent decades there has been a large migration to cities and more than half the world's population now lives in urban areas. Such migration and increasing vulnerability means that even without increasing numbers of extreme events, losses attributable to each event will tend to increase.

The Centre for Research on the Epidemiology of Disasters (CRED) records events where at least 10 people were reported killed; 100 people were reported affected; there was a call for international assistance; or declaration of a state of emergency. There are increasing trends of economic and insured losses from disaster events, and economic annual losses have increased tenfold since the 1950s. However, much of the upward trend in economic losses probably is due to societal shifts and increasing vulnerability to weather and climate extremes.

Some regions are more severely affected than others, although some show a decrease in the number of people killed and the number of people affected.

— increasing concentration of people and property in urban areas

— settlement in exposed or high risk areas

— changes in environmental conditions

There has been an apparent recent increase in the number of disasters but little change in the number of people killed (94). In 2000 there were over 400 disasters, with 250 million people affected. This paradox may be explained by technological advances in the construction of buildings and infrastructure along with advancements in early warning systems, especially in more developed regions. Although there are pronounced year-to-year fluctuations in the numbers of deaths due to disasters, a trend towards increased numbers of deaths and numbers of people affected has been observed in recent decades.

Health Impacts of Natural Disasters

Extreme weather events directly cause death and injury

and have substantial indirect health impacts. These indirect impacts occur as a result of damage to the local infrastructure, population displacement and ecological change. Direct and indirect impacts can lead to impairment of the public health infrastructure, psychological and social effects, and reduced access to health care services. The health impacts of natural disasters include:

— physical injury;

— decreases in nutritional status, especially in children;

— increases in respiratory and diarrhoeal diseases due to crowding of survivors, often with limited shelter and access to potable water;

— impacts on mental health which may be long lasting in some cases;

— increased risk of water-related and infectious diseases due to disruption of water supply or sewage systems, population displacement and overcrowding;

— release and dissemination of dangerous chemicals from storage sites and waste disposal sites into flood waters.

Dangerous floods

Floods are associated with particular dangers to human populations. Immediate effects are largely death and injuries from drowning and being swept against hard objects. Local infrastructure can be affected severely during a natural disaster. El Niño related damage may include: flood damage to buildings and equipment, including materials and supplies; flood damage to roads and transport; problems with drainage and sewerage; and damage to water supply systems.

During and following both catastrophic and non-catastrophic flooding, there is a risk to health if the floodwaters become contaminated with human or animal waste. A study in populations displaced by catastrophic floods in Bangladesh in 1988 found that diarrhoea was the most common illness, followed by respiratory infection. Watery diarrhoea was the most common cause of death for all age groups under 45.

In both rural Bangladesh and Khartoum, Sudan, the proportion of severely malnourished children increased after flooding. In developed countries, both physical and disease risks from flooding are greatly reduced by a well maintained flood control and sanitation infrastructure and public health measures, such as monitoring and surveillance activities to detect and control outbreaks of infectious disease.

TROPICAL CYCLONES AND WINDSTORMS

Impoverished and high-density populations in low-lying and environmentally degraded areas are particularly vulnerable to tropical cyclones, the majority of deaths caused by drowning in the storm surge. Bangladesh has experienced some of the most serious impacts of tropical cyclones this century, due to a combination of meteorological and topographical conditions and the inherent vulnerability of a low-income, poorly resourced population. Improved early warning systems have decreased the impacts in recent years.

Droughts impacts on humans

A drought can be defined as "a period of abnormally dry weather which persists long enough to produce a serious hydrologic imbalance", or as a "period of deficiency of moisture in the soil such that there is inadequate water required for plants, animals and human beings". There are four general types of drought, all which impact on humans, but in different ways:

— meteorological: measured precipitation is unusually low for a particular region;

— agricultural: amount of moisture in the soil is no longer sufficient for crops under cultivation;

— hydrological: surface water and groundwater supplies are below normal;

— socioeconomic: lack of water affects the economic capacity of people to survive, i.e. affects non-agricultural production.

The health impacts on populations occur primarily on food production. Famine often occurs when a preexisting situation of malnutrition worsens: the health consequences of drought include diseases resulting from malnutritions. In addition to adverse environmental conditions political, environmental or economic crises can trigger a collapse in the food marketing systems. The major food emergency in Sudan during 1998 illustrates the interrelationship between climatic triggers of famine and conflict.

8

International Emission Trading

International trade holds the potential of reducing costs of controlling world emissions of green house gases (GHGs) because the nations of the world experience very different costs for achieving emissions reductions on their own. However, the potential gains from trade, like the costs of compliance themselves, may be very unevenly distributed across the world's participants. While all of the parties to an agreement stand to gain collectively under trade in emissions rights as compared with "independent compliance", non-participants in the agreement may either benefit or not depending on their own particular circumstances.

The detailed rules for trading affect how effective trading could be, as well as the level of gains that would be captured in practice. Details of the trading rules will influence both the total gains from trade and distribution of such gains. Key issues include definitions of the emissions rights to be traded, the rules for crediting carbon sinks, and regulations governing participation in the trading framework.

Gains from Trade

The fact that trade produces gains is a powerful point that relates directly to the question of greenhouse gas control.

Most greenhouse gases (GHGs) mix rapidly in the atmosphere, persist for decades or more, and are expected to affect climate. Because GHGs lead to global effects, it does not matter from where GHG reductions come. Countries and regions differ in their degree of dependence on production activities that emit GHGs, the efficiency with which they produce goods and services per ton of GHGs emitted, and the ease with which they can change their current dependency and efficiency.

The principle of gains from trade states that whenever two or more organisations are obligated to produce a fixed amount of a good or service and their marginal costs of production differ, both can be made better off through trade. The gains can be realised if the entity with the higher marginal costs reduces production and pays the entity with the lower marginal costs to increase production.

The principle depends on the difference between marginal costs, not the absolute level of costs. It is equally valid for two low-cost producers or two high-cost producers, so long as costs differ. GHG emissions control would be less costly overall if those countries and organisations that have relatively high costs of emissions reductions were allowed to pay those with lower costs of emissions reduction to undertake more of the actual emissions reductions.

These cost savings would be realised if markets could be established that allowed trading of "permits" or rights to emit GHGs. Nations with higher emissions control costs could then compensate lower-cost nations to undertake emissions control on behalf of the higher-costnations. Of course, the principle is silent on the question of how the savings are actually shared.

Explanation

One country has higher domestic marginal costs of carbon control and the other country has lower domestic marginal costs of carbon control. To make the example as clear and simple as possible, also suppose that both countries have full knowledge of each other's costs so that there is no controversy over where it is least costly to undertake emissions reduction; that both countries have full control over their own emissions and effective access to appropriate control technologies; that in both countries trading of environmental permits is considered an acceptable means of controlling emissions; and, finally, that it costs nothing to specify emissions, transact trades, and enforce compliance.

It makes no difference for purposes of this example whether or not the low-cost controller has any actual obligation to control carbon. Lower overall costs of carbon control would result if, rather than controlling emissions at its own high domestic marginal costs, the high-cost controller compensated the low-cost controller for undertaking emissions control that the low-cost controller was not otherwise obligated to perform.

In a world where both countries had an initial stock of emissions permits corresponding to their annual emissions of carbon, this trade could take the form of the high-cost controller purchasing emissions permits from the low-cost controller. So long as the payment for the permits is greater than the cost of control for the low-cost controller, the low-cost controller would benefit from this arrangement. So long as the payment is less than the costs of control for the high-cost controller, the high-cost controller benefits. So long as the costs of executing, monitoring, regulating, and enforcing trades do not absorb these cost savings, the overall costs of compliance are reduced. These cost savings are the so-called "gains from trade."

In the real world, marginal costs change as mitigation occurs. Every region has a long menu of potential emissions mitigation options. These opportunities range fro m the very inexpensive to the very expensive. Within a region, the cheap mitigation options would be expected to be undertaken before more expensive emissions control options; i.e., there is a schedule of control measures characterised by increasing marginal costs, also known as a "marginal abatement cost curve." If the cost of controlling the last unit of emissions is different between two regions, a potential gain can be shared between the parties by increasing mitigation in the region with the lower marginal cost and decreasing it in the region with the higher marginal cost.

In the real world, of course, significant departures from the idealised circumstances given above would reduce the gains from trade. Economic actors have problems in estimating costs, in accessing and acting upon technological information, in trading, and in enforcing the results of those trades. All of these factors increase the costs of trading and reduce the gains from trade.

ANALYSIS OF CARBON TRADING

The modelling analysis of carbon trading primarily with Pacific Northwest National Laboratory's Second Generation Model (SGM). Similar to several other integrated assessment models, the SGM reflects the recent trend toward hybrid integrated computer models that incorporate features from both energy modelling approaches: the "top-down" approach and the "bottom-up" approach. Economic detail is maintained in the energy supply and transformation sectors that are important for GHG emissions projections, but is aggregated elsewhere into one large "everything else" sector. The SGM, like most integrated assessment models, does a relatively good job of capturing long-run costs, but does not capture transition effects such as inflation rates, unemployment, GNP, and monetary aggregates.

In an equilibrium model like the SGM, markets are linked to other markets through the market clearing process. For example, a change in the demand for coal will have an effect not just on the price of coal, but also on the prices of oil, gas, and-at least indirectly-the prices of all markets in the economy. Thus, in the SGM, markets are said to clear when the model solves for the set of prices for all markets in the modelled economy so that demands and supplies of each market are in balance.

The set of prices in which the equilibrium holds is called the market-clearing price set. The SGM also solves for carbon permit prices as part of this market equilibrium. Specifically, the SGM finds the carbon price at which the amount of carbon abated just satisfies the carbon emissions limitation constraint of a region or group of regions. Thus, SGM provides a consistent way to examine alternative strategies for limiting COand other GHGs and to examine the impacts of energy prices on economic output.

The reductions in carbon emissions that are necessary to return emissions to 1990 levels, or to 10% below 1990 levels, from a base case or "no control" case. It is immediately obvious that, because emissions in each country or region grow over time in the base case due to economic and population growth, the fixed obligation of a return to 1990 emissions levels implies that annual emissions reductions requirements would be greater:

— the faster the rate of economic and population growth,
— the more stringent the level of mitigation that must be reached, and
— the further out in time the emissions controls are implemented.

The Kyoto Protocol 10 targets for Annex I countries 11 range from 10% above to 8% below 1990 levels. The effects of GHG emissions trading on emissions and costs are shown for the years 2010 and 2020. Three control cases are examined in this section. The cases are related to the Kyoto Protocol, under which the so-called Annex I countries have set targets for emissions reductions to be completed during the years of 2008 to 2012. The aim of these cases is not to highlight the costs of implementing Kyoto, but rather to show the magnitude of the costs savings that may occur from international emissions trading.

— *No Trade Scenario*. Each nation is responsible for its own emissions reductions and bears its own abatement costs.

— *Annex I Trading*. Annex I countries are assumed to trade permits among themselves to reduce compliance costs. It is assumed that permits are supplied in competitive markets and that no restrictions exist on the supply or use of permits..

— *World Trading*. This case is introduced to demonstrate the potential gains from trade that could be achieved by having the entire world participate in achieving the Annex I emissions obligation. In this case, the global emissions mitigation limitation remains the same as in the other cases. What changes is the extent to which parties other than those with explicit emissions limitations can participate in the process. This analysis is agnostic about the mechanism by which this extension is accomplished. This case treats non-Annex I countries as if they agreed to distribute permits equal to their annual base case emissions and allowed these permits to be traded internationally. Some would argue that this case corresponds to very broad utilisation of the CDM and perhaps that the inherent limitations of credit trading in general, and of the CDM specifically, would not allow such broad utilisation.

Several basic results that emerge from the analyses conducted with the SGM are listed below.

— Mitigation of carbon emissions will cost less overall if trade in carbon emissions is allowed than if each nation must meet its emissions reduction targets on its own. The broader the trade possibilities, the less the overall costs of control.

— Carbon permit buyers generally will benefit from emissions trading as a method to meet reduction goals. The permit buyers can gain from lower-cost emissions mitigation.

— Given that trading is allowed, potential permit sellers can gain by selling permits whose value exceeds the extra cost of their emissions mitigation. If trading is not allowed, some potential non-Annex I permit sellers could still be better off because Annex I's emissions mitigation efforts will lower the price of remaining energy supplies on the world market.

— Independent compliance by Annex I countries lowers international fossil fuel prices and increases the costs of energy-intensive activities in the Annex I countries. Thus energy—intensive countries could benefit if: (a) they are potential permit sellers, (b) they compete with Annex I countries, and © their export product markets are not too tied to Annex I countries.

— Because the costs of fuels could be affected by emissions control and emissions trading, countries and regions may be affected whether or not they participate in emissions reduction. For example, if emissions trading is forbidden, then relative to the base case: (a) fossil fuel prices fall because the requirement to control carbon reduces fossil fuel demand in Annex I countries as inputs to production; and (b) Annex I economies import relatively more energy—intensive goods but © overall Annex I demand for goods and services falls. The first two effects are positive for non-Annex I countries, but the third is not. The final result depends on which of these effects is the most important within each region. By allowing Annex I countries to reduce their carbon emissions while consuming more energy than they would in the no trading case, trading leaves world oil prices higher, and therefore reduces both the positive and negative effects on non-participants.

No Trade Scenario

Each region's marginal abatement cost is different. The effects of reducing emissions to 1990 levels in 2010 for the No Trade scenario, the carbon reduction required in each region, the marginal costs of abatement, and total costs. Non-participating countries can benefit from "leakage," in which carbon-emitting activity that is constrained in the OECD migrates to other countries that a re not constrained.

In effect, the Annex I countries face an economic penalty for using fossil fuels, while non-Annex I countries face no such penalty. Relative to the base case, Annex I fossil fuel-intensive economic activity becomes less profitable and declines, reducing Annex I demand for fossil fuels, putting downward pressure on world fossil fuel prices, and shifting some fossil fuel-intensive economic activity to non-Annex I countries where it is relatively more profitable. Although non-Annex I countries could absorb some of the residual supplies in the world energy market, most models show some net decline in world fossil fuel prices.

Relative to the base case, the principal economic beneficiaries in the No Trade case are those countries that use large quantities of fossil fuels and that do not rely extensively on OECD markets. However, the direction and level of economic impact on a given country depends on how much international fossil fuel prices are affected by carbon mitigation.

(1) a base case in which oil prices are unaffected by carbon abatement;

(2) a case in which world oil prices fall by 10% relative to their base value as a result of carbon mitigation's negative impact on the world demand for fossil fuels; and

(3) a case in which they fall 20% relative to their base value.

National Gross Domestic Products (GDPs) are lower because of the cost of carbon mitigation to these economies. The off setting effect of lower world oil prices is relatively modest in these countries because end users only see the higher domestic fuel prices that include the embedded marginal cost of carbon reduction. If the world oil price were lowered substantially by carbon mitigation in Annex I countries, however, several of the non-Annex I countries stand to benefit from leakage of energy-intensive economic activity and from lower oil prices. Thus, Korea and India.

The effect on China and Mexico would be more modest, as China restricts oil imports and Mexico is an oil exporting country. Because world oil prices are likely to be higher in a regime with carbon reduction and trading than in a regime with carbon reduction and no trading, leakage would be less with trading than without it.

Role of Annex I Trading

The economic gains from trade are substantial. The effect on carbon permit prices is substantial as well. The United States net permit purchases are 75 million tons of carbon. Thus, it satisfies its 462 million ton obligation in the following manner: 84% from domestic sources and 16%

with purchased permits. The corresponding domestic abatement and purchased permit percentages in Western Europe are 77% and 23% and in Japan, 48% and 52%, respectively. Total costs now not only include the amount spent on domestic emissions control, but also the amount spent on purchasing permits as a substitute for domestic emissions control.

The gains from trading permits among the regions in this set of scenarios are about $20 billion (1992$) worldwide in the year 2010. This reduction in direct cost is 30% of the cost that would have been incurred by the Annex I countries in the absence of an ability to trade permits internationally. The cost of returning emissions to 1990 levels is met entirely within the Annex I countries in this case; there is no obligation on the part of the rest of the world to reduce GHG emissions.

Role of World Trading

The gains from trade are potentially much greater if the group of nations undertaking reductions could be expanded to include the non-Annex I countries as well as the Annex I countries. Although under the Kyoto Protocol non-Annex I countries currently have no obligation to control GHG emissions, this hypothetical case treats non-Annex I countries as if they agreed to create permits equal to their annual base case emissions and allowed these permits to be traded internationally. The non-Annex I countries are expected to grow quite rapidly economically and are expected to burn a considerable amount of fossil fuel with relatively inefficient technology. The decline in overall cost is the result of engaging the world economy in the search for emissions abatement opportunities. Despite the fact that non-Annex I nations have no emissions mitigation obligation, in this case they can search for low-cost abatement opportunities, create an excess in emissions permits relative to their emissions, and sell the excess permits at the world price. This reduces the pressure to prematurely retire existing capital stock in the developed world, and allows a greater share of emissions abatement to come from altering the character of new investments at the lowest marginal cost.

The broader trading possibilities mean that the marginal abatement cost is much lower than it otherwise would be with more limited trading opportunities. A larger number of regions now participate in GHG mitigation. The overall gains from trade are $49 billion with world trading, an improvement of $29 billion relative to Annex I trading. The broader trading regime takes advantage of the more abundant abatement opportunities and greater disparity of marginal abatement costs among the Annex I and non- Annex I countries to achieve greater cost savings.

The permit buyers like the United States and other OECD countries "win" because they achieve their emissions obligations at much lower total direct cost than if they had to achieve all of their abatement domestically, or even if they could only trade emissions permits with other Annex I countries. The Annex I countries benefit because, as a group, the direct cost of fulfilling their carbon obligations is reduced by $41 billion relative to the No Trade case. The Former Soviet Union and Eastern Europe would still benefit from world trading; however, they are not as well off as they would be under Annex I trading, when they were the only net suppliers of permits. Non-Annex I regions on balance benefit relative to the No Trade case from undertaking emissions reductions and then selling their permits to the Annex I countries.

These calculations estimate the potential value of extending participation to the entire world. Real world gains from trade will likely be smaller. Costs of monitoring and compliance will certainly increase as trade expands from the narrow domain of Annex I nations with extensive monitoring and verification capabilities to encompass the entire world.

If, as in the case of the Kyoto Protocol, non-Annex I nations have no formal emissions limitation obligation, mechanisms such as the CDM will have to be used to approximate the case modelled here. With a second-best policy instrument such as the CDM, either the supply of credits and/or the actual environmental benefit will be smaller than what is estimated for a perfect world.

Models of Carbon Trading

Several modelling groups have undertaken empirical analyses to estimate the impact of carbon trading on emissions and costs of GHG abatement, and all have projected substantial economic benefits.. While there are quantitative differences among models due to reasonable differences in assumptions and differences in the details of model structure, there is very broad agreement in recent modelling analyses conducted around the world that emissions trading could substantially reduce the costs of accomplishing any given level of carbon emissions reduction. Many of these analyses were either performed in anticipation of the international agreement reached in Kyoto, Japan in 1997 or they were performed after the fact in an attempt to understand the implications of that agreement.

If it enters into force, the agreement would require specific cuts in emissions of six GHGs. Annex I signatories agreed to adopt national policies to return anthropogenic emissions of GHGs to levels averaging approximately 5.2% below 1990 levels during the years 2008 to 2012, with average OECD emissions reduced approximately 7%. The most important of these arises as a consequence of the allocation of emissions to the FSU and Eastern Europe.

In 2010, allowed emissions in these two regions may be greater than reference case emissions. Emissions permits are distributed based on 1990 emissions. Therefore, total Annex I emissions abatement without trade could be greater than the abatement with trade in the first compliance period. In the results reported in this section, under the Kyoto Protocol, cost reductions occur for two reasons: (1) "where" flexibility, and (2) a reduction in the required net Annex I abatement in the first compliance period. Results reported below do not distinguish between these two causes of gains to trade.

In the long term, of course, cumulative emissions would be the same with and without trading. In a "no trade" case, the excess of allowable emissions over reference emissions in the Former Soviet Union's first budget period would be banked and utilised in subsequent budget periods. Thus, a trading case would have lower emissions in the future relative to a "no trade" case. From the perspective of the environment, the difference between the two profiles is that the "no trade" case has a slightly lower near- term GHG concentration and a slightly higher long-term GHG concentration.

Use of Models

The comparative analyses in this section uses the results from eight models. The models include the Pacific Northwest National Laboratories SGM; the Massachusetts Institute of Technology's Emissions Prediction and Policy Assessment (EPPA) Model; the MERGE model of the Electric Power Research Institute (EPRI); the National Energy Modelling System (NEMS) of the U.S. Energy Information Administration ;Charles River Associates' International Impact Assessment Model (IIAM); the OECD GREEN model from the OECD Development Centre in Paris; the Global Trade and Environment Model (GTEM) of the ABARE modelling group in Australia; and the G-Cubed model by Mc Kibbin, Shackelton and Wilcoxen.

There are many similarities between the models used in this comparison. Except for the NEMS analysis, all of the projections in this section came from multiregional economic models that feature international trade in goods and services. These projections allow for examination of the effects of actions taken in one region on the economies in other regions. These models are hybrid computer models of the economy and energy sector that belong to a class of models known as computable general equilibrium models.

The main apparent differences occur here because of the various baselines assumed by the different modellers. For example, the modellers disagree on such assumptions as the elasticity of energy demand and supply technologies to fossil fuel prices and future economic growth rates in specific regions. The EPPA modellers, for example, appear to be sanguine about Eastern Europe's future economic growth prospects, so EPPA projects that reductions would be required in Eastern Europe in 2010, while the other models do not.

While different modellers have assumed slightly different levels of economic activity, in general most of the analyses are close enough to provide a useful source of comparison. While the results differ in detail, the main empirical findings in Section III hold up.

— Permit buyers benefit from lower compliance costs and permit sellers benefit from being paid more for permits than their additional mitigation costs. If trading is allowed, potential permit sellers will be better off if they trade than if they do not.

— Oil exporters benefit from trade in permits because oil prices are generally higher than if each nation independently met its own mitigation obligations. This occurs both because trade shifts emissions abatement away from oil and toward coal, and because allowing trade raises Annex I total emissions in the first budget period. Results for other non-participants are mixed.

— Annex I parties are richer than they would be without trade, and their increased income enhances the general demand for imports of goods and services from non-participants. On the other hand, individual countries may be more affected by specific changes in relative competitiveness in specific industries and by individual fuel price changes.

Trade by Buyers and Sellers of Permits

Because the underlying assumptions and model structures differ, so do the marginal abatement costs, net mitigation costs and effects on GDP. However, all of the models show substantial savings from trade. Marginal abatement costs are generally about 18% to 50% lower than without trade, net costs 15% to 75% lower, and GDP losses 0% to 2.2% lower.

Non-Participants and Leakage

Emissions trading could also significantly affect the prices of fossil fuels and trade in both these fuels and in other goods, which in turn affects the economies of non-participants. Key effects on non-participants have been examined in many of the models reporting international results. One significant consequence of carbon control is that carbon intensive economic activity migrates from Annex I countries to non-Annex I countries. The EPPA model also reports a worldwide fall in oil prices and natural gas prices which negatively affects the revenues of the regions that export these commodities. These fuels are traded extensively internationally.

Use of gas and oil by the Annex I countries falls by between 3% and 25%, depending on the region, but other regions that are not constrained by the Kyoto Protocol take up most of this consumption. Thus, the main effect on energy—exporting countries is through the decline in energy prices. Output of energy-intensive goods declines in the Kyoto-constrained regions, but expands significantly in the rest of the world. The analysis does not include effects on overall GDP or other broader economic measurements.

The EPPA model also provides some additional information on the effects of carbon permittrading on carbon leakage and non-participant economies. With Annex I trading, the domestic price of coal in non-participant nations falls relative to the base case in those regions where it is used, but the decrease is only about half as large as it is under the No Trade scenario.

In contrast to the No Trade case, the international prices of oil and natural gas are virtually unaffected under Annex I trading relative to the base case. They decrease by less than 1% and 4% respectively, and use of these fuels in the Annex I countries falls by less than 1%. Thus, energy—exporting countries see a much smaller decline in their revenues than when trade is not permitted. With trading, there is also a much smaller decline in the production of energy-intensive goods in the Annex I countries than when trade in permits is prohibited. The rest of the world then shows a decline in the production of energy-intensive goods rather than the major increase shown in the No Trade case.

An analysis of carbon leakage from the Annex I to the non-Annex I world by Charles River Associates (CRA) in January 1997 did not deal with trade of emissions permits, but did point out that without trade, the effects on the non-Annex I world could be significant even though they were not involved in controlling carbon emissions. Generally speaking, non-Annex I countries are affected in one of three ways in the CRA's analysis:

— Trading affects the prices of fossil fuels. In the No Trade case, international oil prices fall substantially relative to the base case. This generally harms the economies of oil export i n g countries. The shift from No Trade to Trade moderates the negative impact on oil prices and oil exporting countries. This occurs both because: (1) emissions trading shifts emissions abatement away from oil and toward coal; and (2) allowing emissions trading raises the Annex I total emissions in the first budget period.

— Non-Annex I countries that are energy-intensive generally benefit from the lower fossil fuel prices and from the increase in demand for energy-intensive goods by Annex I nations.

— Non-Annex I countries that trade mainly with other non-Annex I countries generally would see expanding markets for their products, but would generally see shrinking markets among participant countries. Trade in permits tends to moderate this effect.

Thus, if Annex I permit trading is not allowed, non-Annex I energy-intensive, energy—importing countries that traded mostly with non-Annex I countries would benefit financially from carbon mitigation. Those countries that benefit from permit trading tend to have the opposite characteristics. C R A's extensive analysis of non-Annex I countries in the year 2030 shows that a comparative handful of non-Annex I countries benefit from non-participation in the No Trade case, ranging from Jamaica with 1.5% gain in year 2030 GDP, to Ghana at about 0.1% gain. Losses range from small to significant. Because carbon trading reduces the energy price effects of carbon control, trade would reduce both gains and losses, but would not necessarily change losers to winners.

Models that include a more sophisticated treatment of international financial flows, such as G-Cubed, show additional effects. In such models, carbon mitigation in the No Trade case has a negative impact on rates of return on capital in the Annex I countries relative to the non-Annex I countries. This causes capital outflows to the non-Annex I world. This effect in turn leads to exchange rate appreciation of these countries' currencies relative to the dollar, yen, and other Annex I country currencies.

This strongly limits the non-Annex I countries' advantage in exports of carbon-intensive goods. However, the exchange rate appreciation has two other beneficial effects that are not reflected in their GDP on the inhabitants of the non-Annex I countries. First, exchange rate appreciation means that non-Annex I dollar-denominated international debt is now less expensive, improving their net international investment position. Second, their imports of goods and services from the Annex I countries are also less expensive. Both effects increase domestic wealth in the non-OECD countries.

Initial Allocation of Permits

In several of the models, the initial allocation of emissions permits under the Kyoto Protocol means that the countries encompassing Eastern Europe and the Former Soviet Union will have emissions which are below their emissions limitation in 2010 without any explicit abatement efforts. This is due both to their current and expected poor economic performance and to economic restructuring away from energy—intensive industry. As a consequence, these regions are not expected to regain their 1990 emissions levels by 2010. Poor economic performance and economic re structuring are potentially important sources of emissions abatement within Annex I countries.

Concern that emissions trading under the Kyoto Protocol would lead to higher Annex I emissions in the first budget period than "no trade" has led to the view in some quarters that permits granted to the Former Soviet Union and Eastern Europe should not be tradable. This would prevent environmental benefits from declining under permit trading relative to the no-trade case. The problem with this argument is that it is static. It presumes that there is only one budget period, the period 2008-2012.

On the other hand, if permits could be applied to multiple budget periods and the Former Soviet Union and Eastern Europe were not allowed to trade, they could "bank" excess permits from the period 2008-2012, then use them in subsequent periods when their national emissions exceeded the quantified emissions limitation. These parties' emissions would therefore simply be moved into the future. Over time, cumulative Annex I emissions would there fore be the same whether or not the Former Soviet Union and Eastern Europe were allowed to trade. And to the extent that there is any difference to the environment, the long-term, year 2050 concentration would be somewhat lower if emissions are released earlier in the century rather than later, because the natural removal processes will have had longer to work.

Other Sensitivities

Estimation of the gains from trade discussed in this section is sensitive to a number of key assumptions. These include the effects of non-CO_2 trace gases and carbon sinks on compliance costs. Non-CO_2 Trace Gases. Non-CO_2 GHGs include methane (CH), nitrous oxide (NO), perfluorocarbons (PFCs), hydrofluoro carbons (HFCs), and sulfur hexaflouride (SF). Emissions of all non-COtrace gases are projected to grow substantially unless they are controlled. Multi-gas control has been explicitly examined in both the SGM and EPPA models. Edmonds et al. 1998 looked at three sensitivity cases to bound the effects of non-COtrace gases on direct costs:

— Costs of controlling non-CO_2 trace gases are proportional to the costs of controlling CO_2.

— Non- CO_2 trace gases can be controlled at zero marginal cost, so control of all GHGs costs only as much as controlling CO_2 alone. This places a lower bound on the cost of multiple-gas control.

— Control of non-CO_2 trace gases has an infinite marginal cost, so all commitments must be met from CO alone. The effective CO_2 control target becomes more stringent and places an upper bound on the cost of multiple-gas control.

The EPPA analysis looked at the infinite cost case, but also explicitly examined control of non- CO_2 trace gases using a marginal cost relationship for each gas in each region. If the marginal costs of control for non-CO_2 trace gases are less than those for CO_2, meeting the requirements of the Kyoto Protocol for non-CO_2 trace gases would be less expensive without trade, and trading less attractive for these gases than for CO_2. No explicit analysis was found of the cost effects of trading permits for emissions of these gases.

It is important to note that the inclusion of non-CO_2 trace gases brings with it additional obligations to reduce emissions, not simply low-cost alternatives to CO_2 control. Even so, the Reilly et al. analysis in fully accounting for multiple gases' sources and sinks appears to reduce the overall costs of compliance. The Edmonds et al. SGM analysis does not allow for sinks, which they separately calculate would reduce the independent marginal cost of compliance in the U.S. from $168 to under $120 per ton; it equates the cost of controlling non-CO_2 gases to the cost of controlling only CO_2. Therefore, the SGM appears to show that controlling all gases might be more expensive than controlling CO_2 alone, whereas this might not be the case.

Sinks. Atmospheric GHG concentrations change not only because of emissions due to fossil fuels but also because of changes in terrestrial sources and sinks from changes in land use or agriculture and forestry practices. The Kyoto Protocol provides credit for new "direct human-induced land-use change and forestry activities, limited to afforestation, reforestation, and deforestation since 1990 " - that is, terrestrial carbon sinks established after 1990. Sequestration in soils and other reservoirs is not yet considered. Strict interpretation of Article 3.3 leaves little room for counting sinks toward emissions mitigation in Annex I nations, with the exception of Australia, which has net land-use emissions in 1990. A strict interpretation of Article 3.3 removes an important potential source of net GHG emissions from the accounts.

A full accounting of all net emissions from land use changes could have a significant impact on both marginal and total costs in those cases where a country has significant terrestrial capacity available. In the case of Canada, for example, full allowance for terrestrial carbon sinks could provide a credit equivalent to 80 million tons of carbon emissions, enough to more than satisfy Canada's Kyoto obligations. In the case of the Former Soviet Union and Eastern Europe, sinks offer up to 213 million tons of additional potential baseline credits that could be sold. Overall, Edmonds et al. conclude that full allowance for terrestrial carbon sinks could reduce the Annex I joint trading permit price from $73 to $23 for meeting the goal of emissions 5.2% below 1990 levels. Entering their emissions estimates into the MIT Integrated Global Systems Model (IGSM), which takes account of climate and ecosystem effects as well as natural sources and sinks,

Reilly et al. conclude that achieving the same reduction in warming in the year 2100 by control of only fossil fuel-based CO_2 costs 60% more than if other GHGs and terrestrial sinks are considered.

Institutional Issues

Numerous issues concerning carbon trading regimes have yet to be worked out, but could significantly affect the various parties involved in trading. These issues include:

— whether countries' control regimes will be compatible with trading;

— the effect of restrictions on permit availability or demand;

— impacts of international transfer payments;

— measurement and reporting of emissions, sinks, and costs; and

— accountability and enforceability.

Many of these issues are examined in much greater depth in an upcoming Pew Centre paper on institutional issues and trading. They are of concern here because of their impacts on the effectiveness and cost of trading and on the volume of permits traded.

A key implication of this discussion of institutional issues is:

The actual cost savings from trade in emissions will likely be less than the theoretical savings shown in most analyses performed with integrated assessment models because these models do not include the various measurement, verification, trading, and enforcement costs that would be characteristic of any real trading system.

Control Regimes with Trading

The United Nations Conference on Trade and Development's (UNCTAD's) Greenhouse Gas Emissions Trading Project determined as one of its three basic assumptions that any global GHG trading system would leave each country free to choose its own domestic policy mix for controlling GHG emissions. Domestic controls could take one of three forms : (1) taxes on GHGs, carbon, energy, fossil fuels or fossil-energy—related activities ; (2) command-and-control regulations that directly limit emissions or prescribe certain technologies or activities ; or (3) allocating emissions allowances that can then be traded among emitting entities.

Permit Availability and Permit Demand

Restrictions on permit availability or demand due to regulation or monopolistic market behaviour could reduce the gains from trade that are actually achieved. All of the analysis thus far has assumed that the market for emissions permits functions smoothly and without restrictions. The Ellerman et al. analysis notes that if usable world perm i supply is low relative to its potential, then the world permit price required to meet Kyoto obligations rises from $31/ton of carbon with unconstrained world trading, to $55/ton with 50% availability and to $230/ton with only 5% availability. Edmonds et al. examined this issue in the context of Annex I trading only and showed that the permit price rises from $73/ton with unconstrained Annex I trading to $113/ton if no permits were available from the FSU and Eastern Europe. Thus, the ability to make permits available for trade is critical to the success of any trading programme. Some regions' governments could utilise their position to gain monopoly power in the permit marketplace to limit the supply of permits and increase their price.

It is also possible that the kinds of sources that actually participate in emissions reduction and trading could be limited for reasons of administrative convenience, political expediency or limited technological options, thereby reducing the potential supply of permits. The allocation of permits is extremely important in determining the cost of monitoring and compliance within the system. It is one thing to allocate permits "upstream "-that is, where carbon enters the economy at the point of extraction or import / export-and another to try to allocate permits at the point of combustion.

The former has far fewer parties involved in a programme of universal coverage than the latter. Systems that try to balance the emissions abatement budget on the backs of a subset of downstream economic activities can be very expensive. To illustrate the cost of narrowly focusing the emissions reduction burden, Edmonds et al. showed that electric utilities' marginal cost of emissions mitigation in the United States would rise to more than 250% of the No Trade case as the utility mitigation burden was arbitrarily raised to 70% of the total. Although trading would help reduce the impact of such exemptions and technology limitations, the widest possible pool of potential permit suppliers would clearly be advantageous for reducing costs.

Transfer Payments

While the pattern of trade in emissions depends on the initial allocation of permits, it is likely that there could be substantial transfers of wealth between some countries and regions associated with the trade in permits. The initial allocation and the rules of trade will decide not only the number of permits traded but also whether a given country or region is a net seller or net purchaser of permits. For example, under Kyoto rules, neither the countries with economies in transition nor the non-Annex

I countries would have to reduce carbon emissions in the year 2010. They might not have permits to sell, and they might not be willing to sell in any case. If trade carbon were confined only to the Annex I countries, most modelling groups show that the United States would be a net seller of permito Japan and Europe.

However, if the economies in transition and the non-Annex I countries were allowed to trade carbon permits, the United States would be a net purchaser. One of the consequences, for example, of the United States' desire to purchase large numbers of permits from Eastern Europe and the FSU would be substantial capital flows into those countries from the United States. On the one hand, these flows would provide hard currency reserves necessary to rebuild these economies. On the other hand, the flows probably would strengthen the local currency against the dollar. This would help solidify the local standard of living and make importing easier, but would also make their exports less competitive.

These potential large-scale financial impacts of changes in trade are not treated well in many of the current economic/emissions models. The financial flows involved could be substantial and could require careful handling, particularly for economies such as the FSU, where their magnitude is large compared to other financial flows. While small relative to the total U.S. trade and capital accounts, these amounts still re p resent a net change in the trade deficit roughly equivalent to a purchase of 1 million barrels a day of crude oil at the price of $20 per barrel. On the selling side, under Annex I trading, the FSU would receive $17 billion in 2010, equivalent to 75% of Russia's trade surplus in 1997, or all U.S. lending to Russia between 1990 and 1996. Under world trading, the FSU would receive only about $1 billion, but China would receive $7 billion, and India, over $4 billion.

Transaction Costs and Monitoring, Reporting

Transaction costs for monitoring, reporting, and certification could also limit the gains from trade from emissions trading. A structure for emissions monitoring, reporting, and certification must be specified as part of any carbon control system, with or without trading. Each Party included in Annex I must establish a national system for estimation of sources and removals of GHGs.

The following are some of the major institutional issues. Emissions can be monitored either directly using monitoring devices or indirectly using predictive methods. There is a trade-off between accuracy and cost. For example, continuous stack monitoring provides more accurate measurements but is more costly than occasional air sampling or emissions estimates. Self-reporting and certification by countries may take place at the national level, while actual emissions reductions and sequestration will occur at the project or company level.

Two approaches to the uncertainty created by less-than-perfect monitoring systems a re to limit emissions control and trading only to those gases and sources that can be readily and reliably monitored, or to adjust measured emissions using techniques such as presumptive emissions factors. The "presumptive permits" could then be traded. If emissions control and trading are limited to only those gases that can be measured accurately, the potential gains from trade will also be limited. If a presumptive permits system is used, the actual effectiveness of the system may be compromised. One reason that emissions control and trading under the U.S. acid rain programme has been successful is that verifying emissions is comparatively easy and accurate.

At the other end of the spectrum, the United States Initiative on Joint Implementation (USIJI), a pilot programme that may enable future domestic credit for carbon emissions reduction and sequestration from projects outside of the United States, frequently deals with indirect estimation of the emissions prevented or carbon sequestered. Experience with this programme indicates that in indirect approaches, determining and certifying emissions reduction credits are particularly difficult.

The presence of uncertainty suggests that emissions permit supply curves for CO from stationary sources within Annex I countries would require the addition of relatively minor incremental transaction costs. The supply curves for other sources, for other GHGs, for carbon sinks, and for other countries might have prices that contain a substantial premium beyond actual marginal abatement cost.

Accountability and Enforceability

Accountability and enforceability would also be a problem should information concerning permit validity prove inaccurate. Accountability and enforceability are problems that must be solved in all emissions control systems. In the context of trading emissions permits, there is a specific question concerning whether the buyer or the seller is partly responsible for the integrity or validity of the permit. While trading provides some incentive to obtain accurate information concerning permit validity, the burden of diligence tends to fall most heavily on the party liable for permit validity.

The UNCTAD analysis notes that strict seller liability is preferable because it enhances the standardisation and therefore the tradability of permits. If compliance mechanisms are strong and it is easy to rectify any excess emissions-e.g., by frequent settling of accounts, subtracting emissions allocations in the following period, and adding a penalty-strict seller liability might be all that would be needed.

9

Climate Change and Health

As a human-generated and worldwide process, global climate change is a qualitatively distinct and very significant addition to the spectrum of environmental health hazards encountered by humankind. Historically, environmental health concerns have focused on toxicological or microbiological risks to health from local exposures.

Appreciation of this scale and type of influence on human health entails an ecological perspective. This perspective recognises that the foundations of long-term good health in populations reside in the continued stability and functioning of the biosphere's life-supporting ecological and physical systems. It also brings an appreciation of the complexity of the systems and moves beyond a simplistic, mechanistic, model of environmental health risks to human health.

In simplified diagrammatic fashion, approximate chronological succession of environmental hazards, as societies undergo economic growth and consequent increases in the scale of human activity and environmental impact. Historically, on a local scale, category A hazards have predominated. In the early years of the industrial revolution in Europe much of the environmental hazard was at household and neighbourhood level. In the middle decades of the twentieth century developed countries began to reduce the levels of category B hazards, often via environmental legislation-such as the clean air acts of European and North American countries.

Today, category C hazards are increasing, reflecting the great pressures that human societies collectively exert on the biogeophysical systems of this planet. Carbon dioxide emission is an important example of a category C hazard. Emission rates increased markedly during the twentieth century, as worldwide industrialisation proceeded and land-use patterns changed at an accelerating rate.

The scale of environmental health problems has expanded from household to neighbourhood to community to regional to global level. This requires consideration of the "ecological footprint" and how to curtail its size within the limits of global ecological sustainability.

Folke and colleagues have estimated that the cities around the Baltic Sea require an area of land and sea surface several hundred times larger than the sum of the areas of the cities themselves. This large ecological footprint, typical of modern industrialised societies, comprises the supplies of food, water and raw materials and the environmental "sinks" into which urban-industrial metabolic waste is emptied.

The moral dilemma is clear: a world of six billion cannot live at that privileged level of environmental impact. There simply is not enough world available! A recent study has estimated that human demands on the biosphere have exceeded the world's "biocapacity" since the 1970s, and is currently about 25% beyond the sustainable capacity of Earth.

Further, not only can the actions of one population affect the health of distant populations-as with the environmental dissemination of chlorinated hydrocarbons -but actions today may jeopardise the well-being and health of future generations. There is already in motion a process of sea level rise that will continue for many centuries as the extra heat trapped at Earth's surface by the human-amplified greenhouse effect progressively enters the deep ocean water.

Similarly, it is likely that the continuing rapid extinction of populations and species of plants and animals will leave a biotically impoverished, less ecologically resilient and less productive world for future generations.

Despite global climate change currently being the most widely discussed of various recent global environmental changes, there is mounting evidence that humans, in aggregate, are overloading the planet's great biogeochemical systems. This has been well summarised by Vitousek and colleagues:

> "Human alteration of Earth is substantial and growing. Between one-third and one-half of the land surface has been transformed by human action; the carbon dioxide concentration in the atmosphere has increased by nearly 30% since the beginning of the Industrial Revolution; more atmospheric nitrogen is fixed by humanity than by all natural terrestrial sources combined; more than half of all accessible surface fresh water is put to use by humanity; and about one-quarter of the bird species on Earth have been driven to extinction. By these and other standards."

The long history of climatic fluctuations since the end of the last global glaciation around 15 000 years ago, along with the evidence of recent temperature rises and the IPCC's projected rapid warming in the current century. Several of the rises and falls of great civilisations are shown. Note that the climatic variations before around 1850 essentially were due to natural forcing processes-

cosmological alignments, volcanic activity, solar activity and so on. Since 1850 there has been an increasing influence via human emission of greenhouse gases in excess of the biosphere's capacity to absorb them without an increase in atmospheric concentration.

Recent Scientific Assessments of Climate Change

The latest report from the Intergovernmental Panel on Climate Change (IPCC) makes several compellingly clear points.

1. Human-induced warming has apparently begun: the particular pattern of temperature increase over the past quarter-century has fingerprints that indicate a substantial contribution from the build-up of greenhouse gases due to human activities.
2. A coherent pattern of changes in simple physical and biological systems has become apparent across all continents-the retreat of glaciers, melting of sea ice, thawing of permafrost, earlier egg-laying by birds, polewards extension of insect and plant species, earlier flowering of plants and so on.
3. The anticipated average surface temperature rise this century, within the range of 1.4 to 5.8 °C, would be a faster increase than predicted in the IPCC's previous major report, in 1996. It is the rate of change in temperature that will pose a particular stress upon many ecosystems and species.

The IPCC also reported that even if humankind manages to curb excess greenhouse gas emissions within the next half-century, the world's oceans will continue to rise for up to 1000 years, reflecting the great inertial processes as heat is transferred from surface to deep water. By that time the sea level rise would have approximated 1-2 metres.

The estimated rise in average world temperature over the coming century conceals various important details. Anticipated surface temperature increases would be greater at higher latitudes, greater on land than at sea, and would affect the daily minimum night-time temperatures more than daily maximum temperatures.

Alaska, northern Canada and northern Siberia, for example, could warm by approximately 5°C during the twenty-first century. Indeed, the temperature increases that have occurred already above the Arctic Circle have disrupted polar bear feeding and breeding, the annual migrations of caribou and the network of telephone poles in Alaska.

Global climate change also would cause rainfall patterns to change with increases over the oceans but a reduction over much of the land surface-especially in various low to medium latitude mid-continental regions and in already arid areas in northwest India, the Middle East, northern Africa and parts of central America.

According to glaciologists there is a slight possibility that large sections of the Antarctic ice mass would melt, thus raising sea level by several metres. However, it appears that disintegration did not occur during the warm peak of the last interglacial period around 120000 years ago, when temperatures were 1-2°C higher than now. Nevertheless, substantial melting of Antarctic ice appears to have occurred in a previous interglacial, and several large ice-shelves have disintegrated in the past two decades.

Another possibility is that the northern Atlantic Gulf Stream might weaken and eventually even shut down if increased meltwater from Greenland disturbs the dynamics of that section of the great, slow and tortuous "conveyor belt" circulation that distributes Pacific-equatorial warm water around the world's oceans. Northwest Europe, relative to same-latitude Newfoundland, currently enjoys 5-7°C of free heating from this heat-source. If weakening of the Gulf Stream does occur over the coming century or two, Europe may actually become a little colder even as the rest of the world warms.

Global climate change is only one of a larger set of destabilising large-scale environmental changes that are now under way, each of them reflecting the increasing human domination of the ecosphere.

These include major global changes such as stratospheric ozone depletion, biodiversity loss, worldwide land degradation, freshwater depletion, and others such as the disruption of the elemental cycles of nitrogen and sulphur, and the global dissemination of persistent organic pollutants. All have great consequences for the sustainability of ecological systems: food production; human economic activities and human population health.

There is growing realisation that the sustainability of population health must be a central consideration in the public discourse on how human societies can make the transition to sustainable development. Hence, public, policymakers and other scientists have an increasing interest in hearing from population health researchers, moving towards a view of population health as an ecological entity: an index of the success of longer-term management of social and natural environments.

Indeed this recognition will assist in altering social and economic practices and priorities, to avert or minimise the occurrence of global environmental changes and their adverse impacts.

Change in world climate would influence the functioning of many ecosystems and the biological health of plants and creatures. Likewise, there would be health impacts on human populations, some of which would be beneficial. For example, milder winters would reduce the seasonal wintertime peak in deaths that occurs in temperate countries, while in currently hot regions a further increase in temperatures might reduce the viability of disease-transmitting mosquito populations.

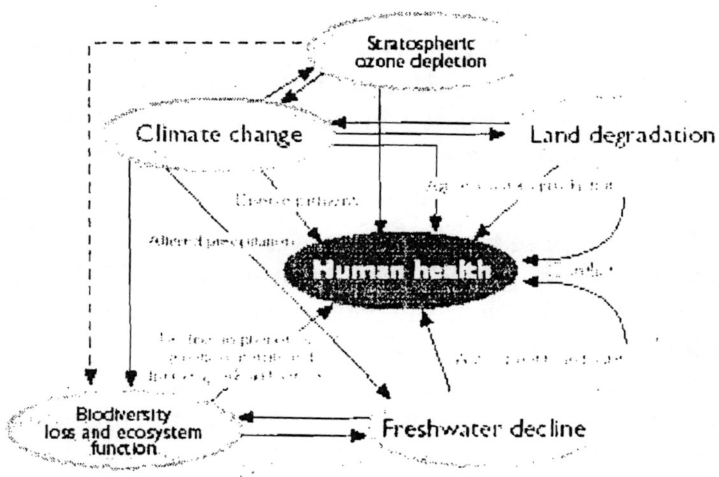

Figure 1. Interrelationships between major types of global environmental change, including climate change. Note that all impinge on human health and—though not shown here explicitly—there are various interactive effects between jointly acting environmental stresses. The diverse pathways by which climate change affects health are the subjects of much of the remainder of this volume.

Climate and Human Health

The Greek physician Hippocrates related epidemics to seasonal weather changes, writing that physicians should have "due regard to the seasons of the year, and the diseases which they produce, and to the states of the wind peculiar to each country and the qualities of its waters".

He exhorts them to take note of "the waters which people use, whether they be marshy and soft, or hard and running from elevated and rocky situations, and then if saltish and unfit for cooking," and to observe "the localities of towns, and of the surrounding country, whether they are low or high, hot or cold, wet or dry... and of the diet and regimen of the inhabitants".

Two thousand years later, Robert Plot, Secretary to the newly-founded Royal Society in England, took weather observations in 1683-84 and noted that if the same observations were made "in many foreign and remote parts at the same time thereby learn to be forewarned certainly of divers emergencies".

Between these times, countless climatic disasters befell communities and pop-ulations around the world, leading variously to starvation, infectious disease, social collapse and the disappearance of whole populations. One such is the mysterious demise of the Viking settlements in Greenland in the fourteenth and fifteenth centuries, as temperatures in and around Europe began to fall.

Established during the Medieval Warm Period during the tenth century AD, these culturally conservative, livestock dependent, settlements could not cope with the progressive deterioration in climate that occurred from the late Middle Ages. Food production declined and food importation became more difficult as sea ice persisted. To compound matters, the native Inuit population in Greenland was pressing southwards, probably in response to the ongoing climate change. The Viking settlements eventually died out or were abandoned in the fourteenth and fifteenth centuries.

Historical accounts abound of acute famine episodes occurring in response to climatic fluctuations. Throughout pre-industrial Europe, diets were marginal over many centuries; the mass of people survived on monotonous diets of vegetables, grain gruel and bread. A particularly dramatic example in Europe was the great medieval famine of 1315-17. Climatic conditions were deteriorating and the cold and soggy conditions led to widespread crop failures, food price rises, hunger and death. Social unrest increased, robberies multiplied and bands of desperate peasants swarmed over the countryside.

Reports of cannibalism abounded from Ireland to the Baltic. Animal diseases proliferated, contributing to the die-off of over half the sheep and oxen in Europe. This tumultuous event and the Black Death which followed thirty years later, are deemed to have contributed to the weakening and dissolution of feudalism in Europe.

Over these and the ensuing centuries, average daily intakes were less than 2000 calories, falling to around 1800 calories in the poorer regions of Europe. This permanent state of dietary insufficiency led to widespread malnutrition, susceptibility to infectious disease and low life expectancy. The superimposed frequent famines inevitably culled the populations, often drastically.

In Tuscany, between the fourteenth and eighteenth centuries there were over 100 years of recorded famine. Meanwhile in China, where the mass rural diet of vegetables and rice accounted for an estimated 98% of caloric intake, between 108 BC and 1910 AD there were famines that involved at least one province in over 90% of years.

Food shortages are never due to climate extremes alone; the risk of famine depends also on many social and political factors. For example, a strong El Niño event in 1877 caused failure of the monsoon rains in south and central India. However, the intense famine that resulted, which caused somewhere between 6 and 10 million deaths, was only partly due to the drought. There was no shortage of food in India at this time, but a large proportion of the Indian population was unable to access food reserves, or to find alternative sources when their usual crops failed. There were many reasons for this. Under the British Raj, common lands that previously provided sustenance in times of hardship had been converted to private property.

Local economies had been impoverished by punitive tariff schemes that favoured imported United Kingdom goods over local products. Aided by the expansion of the railways, community-controlled reserves of food had been replaced by remote stockpiles but there were no moral or regulatory controls over speculation.

Health Impacts of Climate Change

Global climate change would affect human health via pathways of varying complexity, scale and directness and with different timing. Similarly, impacts would vary geographically as a function both of environment and topography and of the vulnerability of the local population.

Impacts would be both positive and negative. This is no surprise since climatic change would disrupt or otherwise alter a large range of natural ecological and physical systems that are an integral part of Earth's life support system. Via climate change humans are contributing to a change in the conditions of life on Earth.

The main pathways and categories of health impact of climate change are shown in Figure 2.

Climate change, acting via less direct mechanisms, would affect the transmission of many infectious diseases and regional food productivity. In the longer term and with considerable variation between populations as a function of geography and vulnerability, these indirect impacts are likely to have greater magnitude than the more direct.

For vector-borne infections, the distribution and abundance of vector organisms and intermediate hosts are affected by various physical and biotic factors. Various integrated modelling studies have forecast that an increase in ambient temperature would cause, worldwide, net increases in the geographical distribution of particular vector organisms although some localised decreases also might occur. Further, temperature related changes in the life-cycle dynamics of both the vector species and the

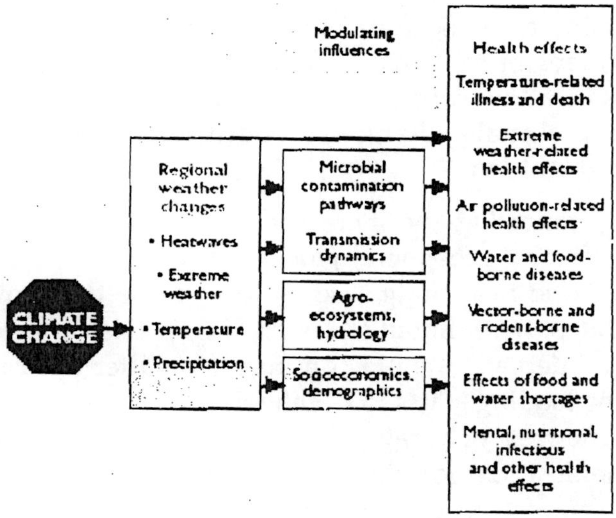

Figure 2. Pathways by which climate change affects human health, including local modulating influences and the feedback influence of adaptation measures.

pathogenic organisms would increase the potential transmission of many vector-borne diseases such as malaria, dengue fever and leishmaniasis -although schistosomiasis may undergo a net decrease in response to climate change.

Recently, there has been considerable effort in developing mathematical models for making such projections. The models in current use have well recognised limitations-but have provided an important start. Allowing for future trends in trade and economic development, modelling studies have been used to estimate the impacts of climate change upon cereal grain yields.

Globally, a slight downturn appears likely but this would be greater in already foodinsecure regions in south Asia, parts of Africa and central America. Such downturns would increase the number of malnourished people by several tens of millions in the world at large-that is, by at least several per cent against a current and projected total, without climate change, of between four and eight hundred million.

By reflecting the increased retention of heat energy in the lower atmosphere, global warming also affects the atmospheric heat budget so as to increase the cooling of the stratosphere. Should this cooling persist, the process of ozone depletion could continue even after chlorine and bromine loading starts to decline. If so, the potential health consequences of stratospheric ozone depletion would become an issue for climate change.

It is likely that climatic change over the past quarter-century has had various incremental impacts on at least some health outcomes. However, the time at which any such health impacts of climate change first become detectable particularly depends upon, firstly, the sensitivity of response and, secondly, whether there is a threshold that results in a "step function". Further, detectability is influenced by the availability of high-quality data and the extent of background variability in the health-related variable under investigation. Detection is a matter of both statistical power and reasonable judgement about attribution.

The former depends on numbers of observations and the extent of divergence between observed and expected rates or magnitudes of health outcomes. The latter includes pattern recognition: if a particular infectious disease undergoes changes in occurrence in multiple geographical locations, each in association with local changes in climate, it is more certain to be due to climatic influence than if such a change occurs in just one setting.

The first detectable changes in human health may well be alterations in the geographical range and seasonality of certain vector-borne infectious diseases. Summertime food-borne infections may show longer-lasting annual peaks. There has been debate, as yet unresolved, over whether recent increases of malaria and dengue in highland regions around the world may be due to climate factors or to the several other factors that are known to be significant determinants of transmission. There are several other categories of likely early impact.

Hot weather would amplify the production of noxious photochemical smog in urban areas and warmer summers would increase the incidence of food poisoning. By contrast, the public health consequences of the disturbance of natural and managed food-producing ecosystems, rising sea levels and population displacement for reasons of physical hazard, land loss, economic disruption and civil strife, may not become evident for several decades.

Vulnerability of Population and Adaptive Responses

Human populations, as with individuals, vary in their

vulnerability to certain health outcomes. A population's vulnerability is a joint function of, first, the extent to which a particular health outcome is sensitive to climate change and, second, the population's capacity to adapt to new climatic conditions. The vulnerability of a population depends on factors such as population density, level of economic development, food availability, income level and distribution, local environmental conditions, preexisting health status and the quality and availability of public health care.

Adaptation refers to actions taken to lessen the impact of climate change. There is a hierarchy of control strategies that can help to protect population health. These strategies are categorised as: administrative or legislative, engineering, personal-behavioural. Legislative or regulatory action can be taken by government, requiring compliance by all or designated classes of persons. Alternatively, adaptive action may be encouraged on a voluntary basis, via advocacy, education or economic incentives.

The former type of action would normally be taken at a supranational, national or community level; the latter would range from supranational to individual levels. Adaptation strategies will be either reactive, in response to climate impacts, or anticipatcry, in order to reduce vulnerability. Adaptation can be undertaken at the international/national, community and individual level- that is, at macro, meso and micro-levels.

The reduction of socioeconomic vulnerability remains a priority. The poor are likely to be at greatest health risk because of their lack of access to material and information resources. Long-term reduction in health inequalities will require income redistribution, full employment, better housing and improved public health infrastructure. There must be improvement in services with a direct impact on health such as primary care, disease control, sanitation and disaster preparedness and relief. The vulnerability of the poor may jeopardise the well-being of more advantaged members of the same population.

Improved environmental management of health-supporting ecosystems would reduce the adverse health impacts of climate change. A good example is the control of water-borne infections. In many areas increased density of rainfall is likely to lead to more frequent occurrence of significant human infections such as giardiasis and cryptosporidiosis.

Traditional public health interventions that focus entirely on personal hygiene and food safety have limited effectiveness. A broader approach would consider the interactions between climate, vegetation, agricultural practices and human activity-and would result in recommendations for the type, time and place of "upstream" public health interventions such as changes in management of water catchment areas.

The maintenance of national public health infrastructure is a crucial element in determining levels of vulnerability and adaptive capacity. The 1990s witnessed the resurgence of several major diseases once thought to have been controlled such as tuberculosis, diphtheria and sexually-transmitted diseases.

The major causes were deteriorating public health infrastructure as well as socioeconomic instability and population movement. Elementary adaptation to climate change can be facilitated by improved monitoring and surveillance systems. Basic indices of population health status are available for most countries.

Such top-down approaches should be widely supplemented by adaptation at the community and indi-vidual levels. These would include local environmental management, urban design, public education, neigh-bourhood alert and assistance schemes, and individual behavioural changes. When implementing adaptation technologies care must be taken to prevent adverse secondary impacts that is, new health hazards created by the application of technologies.

10

Impacts of Climate Change to Coral Reefs

Coral reefs are striking, complex, and important features of the marine environment. Reefs are geologic formations constructed from the accumulated skeletons of limestone-secreting animals and plants. The intimately linked plant-animal communities that create them are representative of an ecosystem that occurs in tropical and subtropical waters across the planet, most commonly in shallow oceanic water, and often close to land.

The natural habitat of coral reefs near the junction of land, sea and air is both varied and variable and is a potentially stressful environment. Reef organisms, have evolved adaptations over hundreds of millions of years to cope with recurring disturbances: damage or destruction, followed by recovery or growth. The major climate change factor that is becoming increasingly important for coral reefs is rising ocean temperatures, which have been implicated in chronic stress and descries epidemics, as well as in the occurrence of mass coral bleaching episodes.

Coral Reefs and Organisms and Communities

Coral reefs, and the organisms and communities that build

and live on them, are widely distributed in shallow tropical and subtropical waters of the world. Coral reefs are unique ecosystems in that they are defined by both biological and geological components. The reef is constructed of limestone secreted as skeletal material by corals and calcareous algae. Reef-building corals are colonial animals that house single-celled microalgae, called zooxanthellae, within their body tissues. This symbiotic relationship benefits both partners: the coral obtains food from the plant photosynthesis, the microalgae benefit from nutrients released as waste by the coral, and the two have complementary effects on carbon dioxide exchange that is believed to account for the rapid rates of skeletal growth.

Coral reefs offer many values to human society and to the health of the biosphere. Reefs support fisheries, and reef structures provide natural breakwaters that protect shorelines, other ecosystems, and human settlements from waves and storms. Humans use reefs and reef products extensively for food, building materials, pharmaceuticals, the aquarium trade, and other products. Due to their grandeur, beauty, and novelty, reefs have become prime tourist destinations and, therefore, economic resources.

Less evident are the multiple "ecosystem services" of coral reefs, such as recycling nutrients and providing food, shelter, and nursery habitat for many other species. Many of these services are related to the geologic and biologic structures that create the spatial complexity necessary for the high biodiversity of reefs. The biodiversity is not all marine; humans, like many seabirds and other air-breathing species, have colonised island and coastal environments formed by coral reef communities.

These two ocean basins have very few reef species in common, and the Indo-Pacific accounts for about 85 percent of the world's reefs and a similar proportion of reef biodiversity. The centre of maximum coral diversity is in the Southeast Asian region. Reef communities vary across large-scale environmental gradients, as reflected in the diversity gradients across both latitude and longitude. At smaller scales, the make-up of reef communities differs according to the degree of land influence, storm frequency, and combination of other local and regional factors.

Reef communities produce limestone, which can be well-preserved in the geologic record. Because reefs and corals have existed in one form or another for hundreds of millions of years, this record provides a wealth of information about organism and ecosystem evolution and Earth's environmental changes through time. Corals have not, however, always been the main reef builders, and comparing ancient with present-day reefs can be confusing, because the terms "coral" and "reef" have been used to refer to a wide variety of things.

Coral Reef Crisis

Coral reefs have declined over the course of human history, culminating in the dramatic increase in coral mortality and reef degradation of the past 20-50 years. This "coral reef crisis" is well-documented and has stimulated numerous publications on the future of coral reefs and their vulnerability to environmental change. The causes of this crisis are a complex mixture of direct human-imposed and climate-related stresses, and include factors such as outbreaks of disease, which have suspected but unproven connections to both human activities and climate factors.

Although the crisis is widespread, individual reefs and even whole regions exhibit considerable variation in both health and responses to stress. The Caribbean region has been particularly hard-hit by problems, many of which are well-studied. Caribbean case studies and inter-ocean contrasts help to illustrate both the consistencies and the variations in coral reef responses to complex environmental changes.

Environmental Changes

Major systematic changes include rising atmospheric concentrations of greenhouse gases (GHGs) that influence the earth's energy budget and climate. In addition, the global phosphorus and nitrogen cycles have accelerated because of artificial fertiliser use and massive changes in land use, the hydrologic cycle has been altered by river damming and water diversion as well as climate change, major natural ecosystems have been altered by fishing, forestry, and agriculture, and the ecological and biogeochemical implications of increased atmospheric CO_2 levels go well beyond the effects on global temperature. Because coral reefs occur near the junction of land, sea, and atmosphere, their natural habitats experience both the marine and terrestrial results of any climatic change and are vulnerable to human activities.

We use "acute" and "chronic" to classify various stress factors, discuss their interactions, and integrate their probable combined effects. Acute stresses are those short-term events that cause rapid damage a reef, while chronic stresses act over longer terms and are generally associated with more gradual environmental degradation.

NONCLIMATIC STRESSES

Effects and Categories of Stresses

A wide variety of environmental factors that are not directly related to changes in the climate system have the potential to stress coral reefs. Reef communities have been described as "disturbance-adapted" ecosystems, but that adaptation is to natural rather than human-enhanced disturbances. Cycles of damage followed by recovery are natural aspects of reef persistence, and coral reefs have been described by Done as a "shifting steady-state mosaic "-a regional population of reef communities that are diverse and changing, but in which all of the important types and components are always represented.

Reef decline, as opposed to change or variation, has two components: the initial damage or mortality and the failure of the ecosystem to recover. Reefs can recover from acute stresses and tolerate chronic stresses, but chronically stressed reefs are far less likely to recover from acute stress. As disturbances become more varied and frequent against a background of deteriorating conditions, components of the original coral reef mosaic are progressively replaced by noncoral organisms. Environmental alteration and climate change need to be considered together to predict the future trajectory of coral reef ecosystems, since both can cause chronic and acute stresses; both also vary across time and space and are likely to have strong interactions.

Earthly Inputs

Contaminant and nutrient loading

The symbiotic relationship between reef-building corals and zooxanthellae allows them to thrive in the nutrient-poor "marine deserts" of the tropical oceans. Addition of nutrients to the water may harm corals directly or make the environment less favourable for reefs by promoting growth of phytoplankton and of seaweeds that compete with corals for space on the reef. Humans have altered the nitrogen (N) and phosphorous (P) cycles at least as extensively as they have the carbon cycle, greatly increasing N and P inputs to the world's oceans and coastal zone. The degree to which rising "background" levels of nutrients in coastal waters are affecting coral reefs is unclear, but heavy nutrient loading can have marked impacts. Unlike nutrients, contaminants are generally human-produced toxic or bioactive materials, which have no natural source or are highly concentrated. They include heavy metals, pesticides and herbicides, solvents, fuels, and other compounds. These materials may be part of complex waste streams discharged to the ocean or adsorbed in sediment. Effects of individual contaminants are difficult to document, trace, or separate from combined effects.

Role of Sediment loading

Sediment deposited onto corals interferes with feeding by the polyps and costs the colonies energy to remove. In the extreme, burial by rapid or prolonged sediment deposition is fatal to corals and other bottom-dwellers. Sediment accumulation also inhibits the establishment of new reefs, because coral communities require hard and stable surfaces. Sediment suspended in the water increases turbidity and reduces available light. Reefs that grow in naturally turbid environments, with organisms that are suited to such conditions, may experience low i.npacts from a moderately increased sediment supply, but sediment loading on reefs that are accustomed to low-sediment conditions imposes significant stress.

Sediment on a coral reef can have two sources: transport of soil particles with freshwater runoff from land, or resuspension of sediment already on the sea floor. Human activities have reduced some sediment sources and increased others. Damming of major rivers has dramatically reduced their sediment discharge to the ocean, but large river outflows represent only a small proportion of the world's coastline and are usually not near reefs.

The socioeconomic forces behind increased sediment runoff are difficult to reverse or control, particularly with rising coastal populations. Once there has been substantial influx of land-derived sediment to a reef region, the "new" sediment often remains in place, is subject to resuspension, and makes the substrate where it accumulates less hospitable to corals. Reefs closest to land masses will probably continue to experience the most intense chronic sediment stress, and the effects will be magnified by climatic change in some areas.

Resource Extraction for Overfishing

Overfishing, the unsustainable fishing or collection of particular organisms, is a global problem with a long history of impacts across the entire marine ecosystem. Removal of plant-eating organisms from a reef upsets the competitive balance between corals and seaweeds, often leading to a fundamental change in the community. The chronic stress of overfishing is often hard to avoid on coral reefs. Although seemingly lush and teeming with life, reef communities generate only small amounts of sustainably harvestable biomass.

Removal of large individuals thus has a disproportionate impact on the species' reproduction. Fishing operations also often have acute destructive effects beyond simple removal of target species. By-catch is often wasted, and damage to other reef organisms and the reef structure itself is common. Blast fishing, widespread in the Southeast Asian region, destroys habitat and is extremely wasteful in terms of incidental kills. Muro-ami ("fishnet") fishers mechanically smash shallow patch reefs and net the fish that are driven out.

Consequences of Coastal Zone Modification and Mining

Human efforts to improve or maintain the coastal zone often have unintended ecological consequences. Dredging, land reclamation, shoreline protection, harbour and runway construction, and other similar activities can have direct, acute impacts by destroying coral reef habitats. The impacts of spoil dumping, sediment resuspension, or local contamination may become chronic and extend well beyond the immediate site. Less apparent is the potential long-term chronic stress imposed by altering patterns of both marine and fresh water movement, as has occurred in south Florida. Reef destruction can also result from deliberate mining of nonliving resources. On atolls, reef islands, and in other coastal areas, sand and reef-rock are the only readily and economically accessible building material. Although healthy reefs produce enough sand to supply reasonable uses, sustainable "harvesting" requires careful attention to where and how the material is removed.

Invasive Species

Introduced or invasive species may be inadvertently brought to a reef through travel, shipping, or careless disposal. The worst incidents of species introductions have occurred in terrestrial, fresh water, or estuarine environments, but there is growing evidence that coral reefs are not immune from the problem. A major concern is that pathogens or parasites will be transported across natural barriers to new and vulnerable host populations.

Climatic Change Stresses

Global climate change imposes interactive chronic and acute stresses, occurring at scales ranging from global to local, on coral reef ecosystems. Gas bubbles preserved in polar ice caps show that atmospheric CO_2 concentrations over the past 400,000 years have oscillated between about 180 and 310 parts per million volume, or ppmv; past temperature and sea-level variations mimic the CO_2 fluctuations, with relatively constant minimum and maximum values. The human-caused increase in atmospheric CO_2 is the near-vertical line at the present day end. Accompanying this CO_2 increase is an observed increase in temperature, and a decrease in pH of the surface ocean. IPCC projections show an even greater departure from geologically recent climates by the end of the present century.

Role of Coral Bleaching

The atmosphere and the ocean have warmed since the end of the 19th century and will continue to warm into the foreseeable future, largely as a result of increasing greenhouse gas concentrations. El Niño-Southern Oscillation (ENSO) events have increased in frequency and intensity over the last few decades. This combination has resulted in a dramatic increase in coral bleaching. "Bleaching" describes the loss of symbiotic algae by the coral or other host. Most of the pigments in the usually colourful corals depend on the presence of these plant cells.

The living tissue of coral animals without algae is translucent, so the white calcium carbonate skeleton shows through, producing a bleached appearance. Bleaching is a general stress response that can be induced in both the field and the laboratory by high or low temperatures, intense light, changes in salinity, or by other physical or chemical stresses. Bleaching is the extreme case of natural variation in algal population density that occurs in many corals.

Three types of bleaching mechanisms are associated with high temperature and/or light: "animalstress bleaching," "algal-stress bleaching," and "physiological bleaching". Although all are important to understanding climate-coral interactions, two are particularly relevant to present concerns: algal-stress bleaching, an acute response to impairment of photosynthesis by high temperature coupled with high light levels; and physiological bleaching, which reflects depleted reserves, reduced tissue biomass, and less capacity to house algae as a result of the added energy demands of sustained above-normal temperatures. A rising baseline in warm-season sea-surface temperatures on coral reefs suggests that physiological bleaching is at least partly to blame in some bleaching events. Such chronic temperature stress may also underlie some less obvious causes of reef decline, such as low rates of sexual reproduction

The temperature threshold for bleaching is not an absolute value, but is relative to other environmental variables and to the duration and severity of the departure from the normal temperature conditions of a reef. Bleaching due to thermal stress is not, therefore, limited to areas of normally high water temperature. Coral bleaching events of greatest concern are acute episodes of high mortality and protracted debilitation of survivors in the form of diminished growth and reproductive rates.

Corals with branching growth forms, rapid growth rates, and thin tissue layers appear to be most sensitive to bleaching, and usually die if seriously bleached. Slow-growing, thick-tissued, massive corals appear to be less sensitive and commonly recover from all but the most extreme episodes. Bleaching thus selectively removes certain species from reefs and can lead to major changes in the geographic distribution of coral species and reef community structures.

Role of Global Warming and Reef Distribution

The global distribution of reef-building corals is limited by annual minimum temperatures of ~18°C (64°F). Although global warming might extend the range of corals into areas that are now too cold, the new area made available by warming will be small, and the countervailing effects of other changes suggest that any geographic expansion of coral reefs will be very minor.

Role of Reduced Calcification Potential

The oceans currently absorb about a third of the anthropogenic CO_2 inputs to the atmosphere, resulting in significant changes in seawater chemistry that affect the ability of reef organisms to calcify. Photosynthesis and respiration by marine organisms also affect seawater CO_2 concentration, but the overwhelming driver of CO_2 concentrations in shallow seawater is the concentration of CO_2 in the overlying atmosphere.

Changes in the CO_2 concentration of seawater through well-known processes of airsea gas exchange alter the pH and the concentrations of carbonate and bicarbonate ions. Surface seawater chemistry adjusts to changes in atmospheric CO_2 concentrations on a time scale of about a year. Projected increases in atmospheric CO_2 may drive a reduction in ocean pH to levels not seen for millions of years.

Many marine organisms use calcium (Ca^{2+}) and carbonate (CO_3^{2-}) ions from seawater to secrete CaCO3 skeletons. Reducing the concentration of either ion can affect the rate of skeletal deposition, but the carbonate ion is much less abundant than calcium, and appears to play a key role in coral calcification calcification. The carbonate ion concentration in surface water will decrease substantially in response to future atmospheric CO_2 increases, reducing the calcification rates of some of the most important $CaCO_3$ producers. These include corals and calcareous algae on coral reefs and planktonic organisms such as coccolithophores and foraminifera in the open ocean.

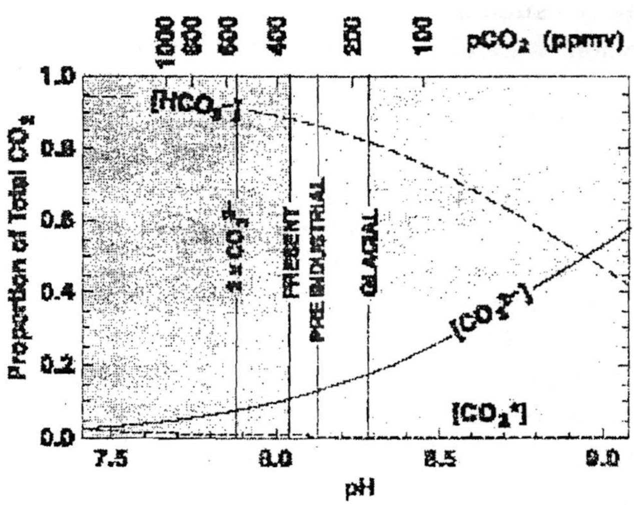

Figure 1 Changes in Carbonate Ion Species

In laboratory experiments that simulate doubled atmospheric CO_2 conditions, coral calcification rates decrease by 11-37 percent, whereas calcareous algae appear to show a stronger reduction of 16-44 percent. Calcification of coral reef communities reflects whether the community is dominated by corals or calcareous algae: the Biosphere coral reef mesocosm, which is dominated by calcareous algae, showed a 40 percent reduction, while coral-dominated mesocosms showed a 14-21 percent reduction in their calcification rates.

Calcification rates of corals also depend on other factors such as temperature. Kleypas et al. estimated an average decline of reef calcification rates of 6-14 percent as atmospheric CO_2 concentration increased from preindustrial levels to present-day values. However, studies have shown that calcification rates of large heads of the massive coral Porites increased rather than decreased over the latter half of the 20th century. Temperature and calcification rates are correlated, and these corals have so far responded more to increases in water temperature than to decreases in carbonate ion concentration. A lowered calcification rate means that calcifying organisms extend their skeletons more slowly and/or form skeletons of lower density. Lower extension rates reduce the ability of corals to compete for space on a reef. Reduced skeletal density means less resistance to breakage and greater susceptibility to both physical breakdown and bioerosion.

Reef-building occurs where calcium carbonate precipitation exceeds its removal. The structural components of reefs are glued together and made more resistant to physical breakdown by calcium carbonate cements that precipitate within the reef framework, and by the over growth of thin layers of calcareous algae. A reduction in $CaCO_3$ precipitation by whatever means reduces a reef's ability to grow and to withstand erosion. Some slow-growing or weakly cemented reefs may stop accumulating or shrink as carbonate deposition declines and/or erosion increases. Such effects have been observed in the Galá pagos and elsewhere.

Role of Sea Level

The predicted rise of sea level due to the combined effects of thermal expansion of ocean water and the addition of water from melting icecaps and glaciers is between 0.1 and 0.9 meter by the end of this century. Sea level has remained fairly stable for the last few thousand years, and many reefs have grown to the point where they are sea-level-limited, with restricted water circulation and little or no potential for upward growth. A modest sea-level rise would therefore be beneficial to such reefs. Although sea-level rise might "drown" reefs that are near their lower depth limit by decreasing available light, the projected rate and magnitude of sea-level rise are well within the ability of most reefs to keep up. A more likely source of stress from sea-level rise would be sedimentation due to increased erosion of shorelines.

Role of El Niño-Southern Oscillation (ENSO)

Mass bleaching of corals in the past two decades has been clearly linked to El Niño events. For example, widespread bleaching events occurred during the El Niños of 1982-83, 1987-88, and 1997-98. During a typical El Niño event, regions of unusually warm water develop throughout the Pacific and Indian Oceans. When these warm water anomalies coincide with seasonal maximum water temperatures, coral bleaching is very likely.

Mean sea level decreases in the western Pacific during an El Niño event, which can expose shallow reefs, and lead to mass mortalities. Coral bleaching has also occurred during the cold phase of ENSO (La Niña) in regions that tend to have warmer-than-normal SSTs.El Niño events have increased in frequency, severity, and duration since the 1970s, but longer-term records do not support a link to global warming. Most global climate models do not predict significant changes in El Niño through the present century, although they do suggest an evolution toward a more "El Niño-like" state.

Role of Ocean Circulation Changes

Circulation, from local to global scales, is likely to change with global climate. Virtually all coral reefs at high latitudes occur where boundary currents deliver warm waters from tropical regions. Changes in the path or strength of these currents would impose different temperature regimes on these reefs. There has been concern that ocean thermohaline circulation (THC) will shut down in the future due to changes in ocean temperature and freshwater runoff. Recent modelling predicts a 0-40 percent slowing of THC within this century, but most models do not predict a complete shutdown. A slowing of THC would lead to significant changes in oceanic circulation and up welling patterns that could potentially affect coral reef ecosystems, but how THC will be affected by global climate change remains uncertain.

Role of Precipitation and Storm Patterns

Tropical precipitation has increased over the past century by 0.2-0.3 percent per decade in the 10°S-10°N region, and the frequency of intense rainfall events is "very likely" to increase over most areas. Increases in precipitation can lower salinity and increase sediment discharge and deposition near river mouths, sometimes leading to mass mortalities on nearby coral reefs. The frequency and intensity of droughts are also expected to increase, which may cause changes in vegetation cover and land use that lead to erosion and sediment stress when rains return.

Tropical cyclones are a fact of life in many tropical regions, and although they may limit reef development in a few instances, healthy coral reefs tend to recover from the infrequent damage caused by cyclones. Comprehensive observations of cyclone activity are limited to the last five or six decades, and these show few trends in cyclone frequency or intensity. Hurricane models and thermodynamic calculations indicate that the maximum potential intensity of cyclones could increase 10-20 percent and surface winds could increase by about 3-10 percent. There is little evidence that the frequency of cyclones or where they form is likely to change.

Combination and Controversy

The coral reef crisis is a global composite of various

interactive problems and variable responses and cannot be understood, predicted, or mitigated on the basis of separate, individual categories of stresses. The effects of changes in predator/prey relationships, and especially coral disease, do not fit conveniently into either the climate change or human stress categories, and they therefore provide an opportunity for considering interactions and synergism. Outbreaks of disease or intense attacks upon corals by predators can occur naturally, but they can also result wholly or in part from climate change or human interventions that modify the natural balance of ecosystems. Recent outbreaks of disease and predation on reefs have been attributed to both human activities and climate change.

Diseases

Disease outbreaks and consequent mortality among corals and other reef organisms have been a major cause of the recent increase in coral reef degradation. Although diseases and syndromes of corals and other reef organisms remain incompletely characterised, they are known to be caused by both bacterial and fungal agents. These diseases are commonly lethal, but they exhibit a wide range of rates of progression. Most appear to affect some species more than others, but few, if any, are species-specific.

Two specific outbreaks have radically altered the ecology of Caribbean coral reefs. One disease killed more than 97 percent of the black-spined sea urchin, Diadema antillarum, some populations of which are now beginning to recover. This sea urchin is an herbivore, and its removal contributed substantially to the transition from coral-dominated to seaweed-dominated surfaces on Caribbean reefs. Another disease, white band disease (WBD), has killed much of the elkhorn and staghorn coral throughout the Caribbean. These were dominant reef-building corals in the Caribbean for tens of thousands to hundreds of thousands of years, but since 1972 WBD has helped reduce these species to candidacy for listing under the Endangered Species Act. Cores taken from the Belizean barrier reef show that A. cervicornis dominated this coral reef community continuously for at least 3,000 years, but was killed by WBD and replaced by another species after 1986. WBD caused a similar, unprecedented change in community structure in the Bahamas.

Effect of Predation

Although coral diseases appear to be increasing in the Indo-Pacific region, a more significant biological stress in parts of the western Pacific has been outbreaks of the crown-of-thorns starfish, an intense coral predator. These outbreaks may be natural events in the crown-of-thorns starfish population cycle, perhaps associated with river floods in wet years, but it has been argued that overfishing of starfish predators or elevated nutrients in freshwater runoff have aided the survival of different life stages of the animal. Like bleaching, A canthaster outbreaks are acute stresses on reefs and probably result from a combination of human and natural factors. Mortality and recovery rates vary among coral species, resulting in shifts in community structure.

Global Climate Change and Human Activity

Climatic warming can increase the virulence of pathogens, since optimal water temperatures for those infectious agents for which data are available are at least 1°C (2°F) higher than the optima of their coral hosts. Recent increases in the frequency and virulence of disease outbreaks on coral reefs are consistent with this prediction, suggesting that the trend of increasing disease will continue and strengthen as global temperatures increase.

Elevated temperatures thus can enhance disease effects by both strengthening the pathogen and weakening the host, and can inhibit recovery by suppressing reproduction. This provides a strong example of potential synergy-and of the difficulty in diagnosing it conclusively.

Two other possible climatic connections to increased coral disease have been proposed. Dust transported from Africa to the Caribbean is a possible source of pathogens, nutrients, and contaminants. Also, land-derived flood plumes from major storms transport materials from the Central American mainland to reefs that are normally considered remote from such influences. Although dust transport and storms are not new phenomena, they now may be acting as vectors for a greater variety of contaminants.

Human actions can also spread disease to coral reefs. Increasing sediment and waste input to many coastal areas enhances the potential for introducing pathogens to coral reefs, and the alteration of nutrient and salinity regimes of coastal waters by human activity may facilitate the propagation of pathogens. Human involvement is suspected in the spread of the pathogen that killed Caribbean Diadema; the disease began in Panama, suggesting a possible link to shipping through the Panama Canal. Travel, shipping, and the importation of marine species for food or the aquarium trade all have the potential to cross-contaminate regions with pathogens not previously present.

Territorial Comparison

The structure and biodiversity of coral reef communities vary from one region to another. The vast, environmentally and biologically diverse Indo-Pacific contrasts markedly with the much more compact and less diverse Caribbean. The Indo-Pacific has approximately 750 species of reef-building corals with 170 in the genus Acropora, compared to about 50 coral species and only two bona fide species of Acropora in the Caribbean. The Caribbean has been recognised as a potentially vulnerable region for these and numerous other reasons.

Recovery times for reefs severely affected by storms and A canthaster vary from a decade to centuries, depending on the frequency of disturbance and the growth and recruitment rates of the affected corals. Caribbean reefs generally recover more slowly than Indo-Pacific reefs. This may reflect the lower success rate of sexual reproduction in Caribbean corals, or the small size of the Caribbean; where disturbances have regional-scale effects, availability of replacement larvae from neighbouring reefs upstream will be limited.

If Caribbean reef-building corals are unable to recover from their currently depressed state, reduced herbivory, and possibly nutrient loading, will prolong the ability of seaweeds to monopolise space. Under such circumstances, bioerosion-the breakdown of reef framework by limestone consuming sponges, clams, and other organisms-and dissolution will exceed the buildup of reef mass.

Vertical and lateral reef growth has probably already slowed or possibly changed to net shrinkage, resulting in a reduction in the extent and variety of reef community habitats. In the Indo-Pacific, the reefs of Southeast Asia have been degraded by pollution, sediment-laden runoff from deforested land, destructive fishing practices, including explosives and cyanide, and other human impacts. As in the Caribbean, the added effects of climate change compound these local stresses.

Effect of Adaptation

Populations and pigment concentrations of symbiotic algae

within the coral tissues adjust to accommodate a wide range of light levels. Corals and reef communities in areas such as the Arabian Gulf tolerate salinity and temperature conditions in excess of any average near-term climate change predictions, conditions that would be lethal if imposed rapidly on the same species in more equable environments. The pressing question is not, "can corals adapt?" but rather, "how fast and to what extent can they adapt?" The recent increases in the frequency and intensity of conditions that contribute to bleaching may have outpaced the compensating mechanisms of many corals.

Over the past 5-10 years, evidence for the diversity of zooxanthellae and environmentally correlated coral-algal partnerships has expanded rapidly, and experiments have shown that the processes required for adaptation driven by bleaching occur in nature. Buddemeier et al. review the evidence and conclude that adaptive bleaching is real, but its operational significance will not be fully known until we have a better understanding of the detailed mechanisms and of the functional taxonomy of the zooxanthellae.

Field data indicate that coral bleaching on some eastern Pacific reefs was much worse during the 1982-83 El Niño than in 1997-98, although temperature extremes during the two events were similar. The difference in responses to these two comparable events offers some support for the idea that corals or communities can adapt to higher temperatures over decades, either through adaptive bleaching or through evolutionary selection for more heat/irradiance-tolerant corals that survive bleaching events.

Risk for Resources

Socioeconomic Evaluation

While estimating direct economic benefits from fishing

and tourism is relatively straightforward, estimating values of services such as shoreline protection, biodiversity, and aesthetic value is not, and these services are often omitted from reef valuations. Two recently completed socioeconomic studies of U.S. reef areas differed in their accounting: a study of reefs of southeast Florida assessed spending and incomes related to reef use, while a study of Hawaiian reefs focused on economic benefits from fisheries, the aquarium trade, tourism, and property values.

The reef-related economic contributions to four Florida counties totalled US$4. billion in sales and $2 billion in annual income. Hawaiian reefs produced an estimated total annual economic benefit of US$363 million. Cesar et al. estimated the global net economic benefit from reefs to be about US$30 billion/year.

Nearly 40 percent of the people on Earth live within 100 km (60 mi) of the coastline, and many local and regional economies are based on goods and services provided by coastal ecosystems. Environmental degradation both on land and in the ocean reduces the ability of reefs to support local economies. This economic loss can lead to even more destructive methods to extract increasingly scarce resources from the reef and adjacent environments, just as the pressures of climate change may cause even more unsustainable land use.

In turn, decreased socioeconomic value of the reef reduces the standard of living of the society that depends on it. Similarly, unmanaged increases in scuba diving and other tourist-related activities on reefs can lead to degradation of the very environment that attracts the tourists.

Biological Impacts

Coral reefs, which support more biodiversity than any

other marine ecosystem, also alter water energy and circulation in many near-shore environments. This shapes other habitats and protects them from wave impact and coastal erosion. Mangrove systems, for example, often develop in quiet near-shore environments protected by reefs and are highly productive nurseries for many important marine species. Loss of reefs as both biological and structural entities would impoverish the marine biota and potentially reduce the large-scale resilience of tropical and subtropical marine ecosystems.

11

Climate Change and Adaption

Several definitions of climate-related adaptation can be found in the literature and continue to evolve. Many definitions focus on human actions, some include current climate variability and extreme events, others are limited to adverse consequences of climate change. The Intergovernmental Panel on Climate Change (IPCC) has developed definitions of adaptation and the closely related concept of adaptive capacity as follows:

— *Adaptation*: adjustment in natural or human systems in response to actual or expected climatic stimuli or their effects, which moderates harm or exploits beneficial opportunities.

— *Adaptive Capacity*: the ability of a system to adjust to climate change to moderate potential damages, to take advantage of opportunities, or to cope with the consequences.

These definitions are comprehensive in that they are not limited to either human or natural systems: both current and future changes in climate are encompassed, and beneficial as well as adverse effects of climate change are included.

Vulnerability Assessment and Adaptation, Climate Impacts

In order to assess health impacts of, and vulnerability to, climate change and variability it is essential to consider adaptation. The ultimate objective of the United Nations Framework Convention on Climate Change (UNFCCC) is "to achieve stabilisation of atmospheric concentrations of greenhouse gases at levels that would prevent dangerous anthropogenic interference with the climate system...". However, the UNFCCC does not define dangerous levels, although it does refer to levels that "allow ecosystems to adapt, ensure food production is not threatened, and enable economic development to proceed in a sustainable manner".

As human population health also depends on these factors, it can serve as an important integrating index of effects of climate change on ecosystems, food supplies, and social-economic development. The extent to which the health of human populations is vulnerable or in danger depends on the direct and indirect exposures of human populations to climate change effects; the populations' sensitivity to the exposure; and the affected systems' ability to adapt. To assess the human health risks associated with climate change, impact and vulnerability assessments must address adaptation.

Adaptation is considered both in the assessment of impacts and vulnerabilities and as a response option. Due to the past accumulation of greenhouse gases, the long lifetimes of these gases and the thermal inertia of the climate system, it is likely that global temperatures will increase and other aspects of climate continue to change regardless of the coordinated international mitigation actions undertaken. Further, it is unlikely that autonomous actions undertaken by individuals or countries in reaction to climate health impacts will fully ameliorate all impacts.

Adaptation and Health Prevention

Many of the adaptive measures discussed in health impact and vulnerability assessments are not unique to climate change. In fact, the IPCC identified rebuilding public health infrastructure as "the most important, cost-effective and urgently needed" adaptation strategy. Other measures endorsed by the IPCC include public health training programmes; more effective surveillance and emergency response systems; and sustainable prevention and control programmes. These measures are familiar to the public health community and needed regardless of whether or not climate changes: they constitute the basis of a "no-regrets" adaptation strategy.

Adaptive actions to reduce health impacts can be considered in terms of the conventional public health categories of primary, secondary, and tertiary prevention. Primary prevention refers to an intervention implemented before there is evidence of disease or injury: avoiding hazardous exposure, removing causative risk factors or protecting individuals so that exposure to the hazard is of no consequence.

Secondary prevention involves intervention implemented after disease has begun, but before it is symptomatic, and subsequent treatment that averts full progression to disease. Examples include enhancing monitoring and surveillance; improving disaster response and recovery; and strengthening the public health system's ability to respond quickly to disease outbreaks. Secondary prevention is analogous to reactive adaptation. Finally, tertiary prevention attempts to minimise the adverse effects of an already present disease or injury.

Climate Variability

Past and current climates have been, and are, variable. This variability is likely to continue with future climate change. In the popular literature, global climate change frequently is called 'global warming' " which focuses attention on average global temperature change. However, a change in climate actually occurs as changes in particular weather conditions, including extremes, in specific places. In many cases the meteorological variables of interest for public health are not averages and may not be confined to temperature alone.

Given current climate variability, climate change adaptations which enhance a country's coping ability can be expected to yield both near-term benefits, as they enable countries to deal better with current variability, and the longer-term benefits of being able to deal better with future climate. Such no-regrets adaptations are likely to be especially important for less developed countries as they result in immediate benefits and are a useful first step in strengthening capacity to deal with future changes.

Many social and economic systems have evolved to accommodate the normal climate and some variation around this norm. This evolution takes place in a dynamic social, economic, technological, biophysical and political context, which determines the coping ability of a region or country.

Coping ability is defined here as the degree to which the public health system and individuals can deal successfully with health effects associated with current climate conditions, including climate variability. It therefore reflects autonomous and planned adaptations that have taken place over time and can be considered the adaptation baseline.

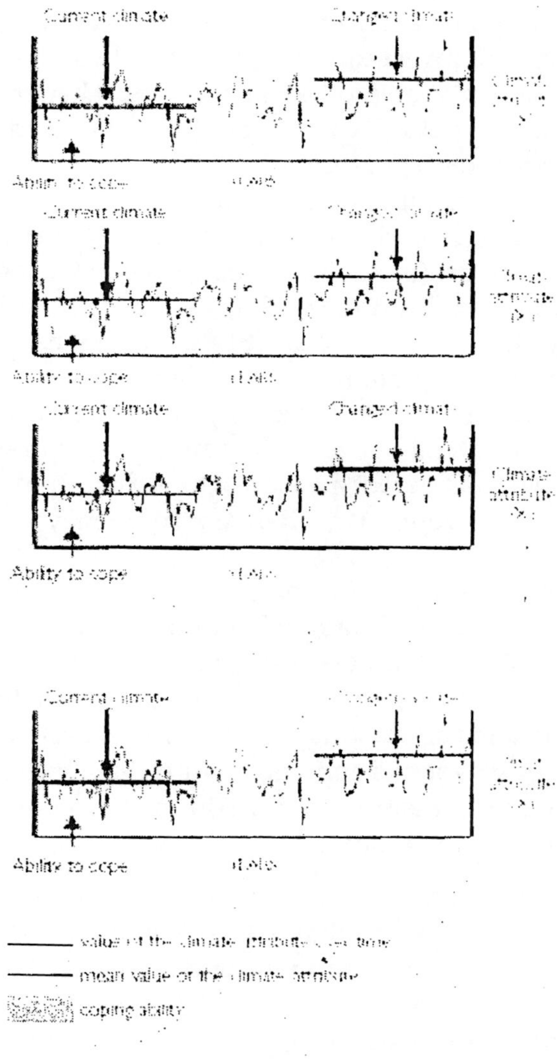

Figure 1 Climate change and coping ability.

Capacity of Adaption

Adaptive capacity encompasses coping ability and

strategies, policies and measures that can expand future coping ability. Adaptive capacity is a theoretical construct because it is not possible to know with certainty whether a country will invest resources to expand its coping ability, how technology and other factors will change, or what adaptations actually will be implemented, until a perturbation or stress occurs.

Decisions about public health measures unrelated to climate change, such as sanitation and water treatment, may have a profound influence on health consequences associated with climate change. In fact, adaptation strategies frequently are described as risk management and public health programmes can be characterised as reducing climate change health risks. Improved weather warning and preparedness systems, buildings and infrastructure, all can be considered measures to reduce human health risks in the event of a changed frequency of weather disasters.

Highly-managed systems, such as agriculture and water resources in developed countries, are thought to be more adaptable than less-managed or natural ecosystems. Similarly, systems that have coped successfully with historical and/or existing stresses are expected to adapt well to stresses associated with future climate change. Both these premises assume that a country's coping ability is maintained or enhanced.

Unfortunately, there are numerous examples in public health where this capability is not maintained once the health threat has been brought under control. Thirty years ago the threat of infectious diseases appeared to be decreasing due to advances in antibiotic drugs, vaccines, and chemical pesticides among other developments.

These types of simple assumptions concerning adaptability have formed the basis for broad assessments of sensitivity and adaptability. Based on these factors, usually it is asserted that much can and will be done to reduce the impacts of climate change. However, it is not clear how much adaptation actually will take place given the number of uncertainties surrounding climate change adaptation. These include uncertainties about future climate, potential effects and underlying determinants of adaptive capacity.

Adaptive Capacity Determination

Research on adaptive capacity in climate change is very limited and is a key research need. However, substantial literature in other fields can provide insights into the likely key determinants of adaptive capacity. These represent conditions that constrain or enhance adaptive capacity and hence the vulnerability of regions, nations and communities. Consideration of these determinants provides another pathway to the overarching goal of protecting and enhancing human health. The IPCC identified the main features of communities or regions that seem to determine their adaptive capacity: economic wealth, technology, information and skills, infrastructure, institutions and equity.

Economic Resources Measures

The economic status of nations, described in terms of GDP, financial capital, wealth, or some other economic measure, clearly is a determinant of adaptive capacity. It is widely accepted that wealthy nations have a greater capacity to adapt because they have the economic resources to invest in adaptive measures and to bear the costs of adaptation. It is also recognised that poverty is related directly to vulnerability and that the poorest groups in the poorest countries are the most vulnerable to health impacts of climate change. Approximately one-fifth of the world's population lives on less than US $1 per day. Excluding the three WHO sub-regions with very low child and adult mortality, a strong gradient of increasing child underweight with increasing absolute poverty was found in the remaining eleven. Unsafe water and sanitation and indoor air pollution also are associated with absolute poverty in these sub-regions.

The feasibility of adaptation options for many poor countries is constrained by a lack of resources. Table 11.1 provides estimates of expenditures on health for the world and by income groups and regions. In 1998 an average of I$ 523 per person was spent on health services. This average varied significantly across both countries and regions, ranging from only I$82 per person in Africa to I$2078 in the OECD countries. Countries with low health expenditures also have poorer health status. The median health-adjusted life expectancy (HALE) in countries that spend less than I$200 per capita on health is 47.1 years.

Income growth, improved educational levels and consequent improvements in nutrition and sanitation have contributed to significant improvements in health and declines in mortality in the twentieth century. The link between economic resources and health can be illustrated further by considering episodes of sharp economic downturns that reduce a country's resources to invest in public health. Combined with poor policy decisions and implementation, adverse economic conditions led to reductions in health expenditures in many developing countries and countries of the former Union of Soviet Socialist Republics in the 1980s.

Research also has begun to reveal the linkage between health and economic growth. While not definitive, this research consistently finds strong relationships between health, as measured by health indicators such as survival rates and life expectancy, and income levels or economic growth rates. Simultaneous impacts of health on wealth and vice versa have been found. Improvements in health affect economic growth directly-" healthier people are more productive" and indirectly, through effects on demography. Research on the effects of geography and climate on income suggests that the interaction of tropical climate and diseases, particularly malaria, can significantly affect economic performance.

Table 1 Health spending in 1998, by income groups and regions.

Income group /region	Total health expenditure (in millions of I$)	Per capita health expenditure (I$)	Share of GDP (%)
World	3072485	523	7.9
Income group			
<1000	9985	24	3.4
1000–2200	156438	88	4.5
2200–7000	518710	206	5.1
>7000	2387353	2042	9.6
Region			
Africa	50170	82	5.1
Americas	176223	438	7.0
Middle East	80932	176	4.8
Eastern Europe and Central Asia	99761	281	5.7
OECD	2317247	2078	9.7
South Asia	141262	95	4.4
Asia and Pacific	206891	142	4.4

Effect of Technology

Advances in technology, such as new drugs or diagnostic equipment, can increase substantially solve health problems. More generally, the availability and access to technology at the individual, local, and national levels, in key sectors is an important determinant of adaptive capacity. Many of the adaptive strategies that protect human health involve technology. While much of this technology is well established, some is relatively new and still being disseminated.

It is important to assess in advance any risks to health from proposed technological adaptations. Increased use of air conditioning would protect against heat stress but could increase emissions of both greenhouse gases and conventional air pollutants.

Similarly, if new pesticides are used to control disease vectors their effects on human health, insect predators, and increased insect resistance to pesticides all need to be considered. New chemicals or treatments for vector control must be effective but their breakdown products should be nontoxic and non-persistent. The migration of potentially hazardous compounds into air and water should be avoided.

Skills and Information

In general, countries with higher levels of "human capital" or knowledge are considered to have greater adaptive capacity. The UN reports that more than 850 million people in developing countries are illiterate and about 90 million children worldwide are denied any schooling, raising concerns about their vulnerability to a range of problems. Illiteracy, as well as poverty, has been listed as a key determinant of low adaptive capacity in northeast Brazil.

As many adaptive measures involve implementation of effective health education programmes, a high level of illiteracy can seriously compromise their effectiveness. Some of the simple, low-cost, low-technology measures to reduce health effects involve educating the public on the feasibility and effectiveness of such measures.

Lack of trained and skilled personnel may restrict a nation's ability to implement adaptation measures. Health systems in particular are labour intensive and require qualified and experienced staff to function well. Health "human capital" can be increased through investment in education and training.

Human capital does not deteriorate with use, but can depreciate as old skills become obsolete with the advent of new knowledge, methods, and technologies. Effective adaptation will require individuals skilled at recognising, reporting and responding to health threats associated with climate change. Researchers trained in epidemiology and laboratory research will be needed to provide a sound basis for surveillance and response. Social scientists can contribute to an understanding of social behaviours and demographics as they relate to causes and control of diseases.

Skilled public health managers, who understand surveillance and diagnostic information, will be needed to mobilise the appropriate response. People trained in the operation, quality control and maintenance of public health infrastructure, including laboratory equipment, communications equipment, and sanitation, waste water, and water supply systems also are required.

Country's Infrastructure

Adaptive capacity is likely to vary with the level of a country's infrastructure. Adaptive responses to health impacts of climate change are enhanced by infrastructures specifically designed to reduce vulnerability to climate variability and general public health infrastructures. Infrastructure such as roads, rails and bridges, water systems and drainage, mass transit and buildings can reduce vulnerability to climate change. It also has the potential to be adversely impacted, which can increase vulnerability to climate change. Flooding can overwhelm sanitation infrastructure and lead to water-related illnesses. After Hurricane Mitch hit Central America, severe damage to the transportation infrastructure made it more difficult to assist affected populations.

Social Institutions

Social institutions are considered an important determinant of adaptive capacity. Those countries with less-effective institutional arrangements, commonly developing nations and those in transition, have a lower capacity to adapt than countries with well-established and effective institutions. Inadequate institutional support frequently is cited as a hindrance to adaptation. Institutional deficiencies and managerial weaknesses are cited as contributing to Bangladesh's vulnerability to climate change.

The Democratic People's Republic of Korea experienced strong storms with torrential rain in 1995 and 1996, followed by droughts in 1997 and 1998. Estimates of deaths from famine since 1995 range from 220000 to 2 million. While there were widespread crop failures, the agricultural system's inability to meet the needs of the people is not new. Other features of this society including economic isolation, lack of reserves, highly centralised arrangements for storage and redistribution of foods, lack of variety in agricultural practice and a strictly hierarchical political system exacerbated the situation.

The IPCC cites "institutional inertia" in the Asia region as limiting investment in environmental protection and increasing climate risks. Inconsistent and unstable agricultural policies have increased the vulnerability of food production in Latin America. Political upheavals in African countries, with the accompanying political and economic instability, constrain the implementation of adaptation measures. Health systems should offer protection against disease, because of economic and social crises these have either collapsed or not been built in many countries.

The complex interaction of issues expected with climate change will require new arrangements and collaborations between institutions to address risks effectively, thereby enhancing adaptive capacity. The Environmental Risk Management Authority in New Zealand involves collaboration between the health, forestry, environment and conservation sectors. Similarly, nations and international organisations such as WHO can cooperate in coordinating surveillance and response activities to address disease threats more effectively.

Effect of Equity

Frequently it is argued that adaptive capacity will be greater if access to resources within a community, nation, or Earth is distributed equitably. Universal access to quality services is a bedrock principle of public health. However, while many have broad and advanced access to health care, many have been denied access. WHO estimates that the developing world carries 90% of the disease burden, yet poorer countries have access to only 10% of the resources for health.

Demographic variables such as age, gender, ethnicity, educational attainment and health often are cited in the literature as related to the ability to cope with risk. WHO notes that for a large group of people long-term unemployment results in exclusion from the mainstream of development and society. The combination of homelessness and lack of access to financial resources and infrastructure restricts adaptation options.

Preexisting Disease Burdens

Population well-being is an important ingredient and

determinant of adaptive capacity. Great progress has been achieved in public health, particularly through the improvement of drinking water and sanitation; development of national health systems; introduction of antibiotics and mass immunisation; and the improvement of nutrition. Yet 170 million children in poor countries are underweight: over 3 million of them die each year as a result. The African region remains the region most affected by infectious and parasitic diseases. Malaria, HIV/AIDS, childhood vaccine preventable diseases and diarrhoea represent the highest estimated deaths in Africa. Malaria is estimated to have caused 963000 deaths and the loss of around 36 million years of "healthy" life in this region in 2001.

Non-communicable diseases, in particular cardiovascular diseases, represent the highest mortality in countries with very low child and adult mortality. However, they are now becoming more prevalent in developing nations, where they create a double burden of disease-a combination of long-established infectious diseases and increasing chronic, noncommunicable diseases.

12

Climate Change Mitigation

The goal of stabilising atmospheric concentrations of GHGs poses a large technological challenge as world energy demand is projected to rise and new technologies will be required to meet this demand. For example, the IEA's World Energy Outlook 2004 estimates that up to US$16 trillion will be invested in the energy sector up to 2030. The IEA estimates that total energy demand, under a business-as-usual scenario, will increase by more that 50 per cent between 2000 and 2030, with developing countries expected to account for 70 per cent of the growth in global energy demand and two-thirds of the growth in global GHG emissions. Fossil fuels are expected to remain the primary source of energy and will meet more than 90 per cent of the projected increase in demand.

Technology for Mitigation

" Mitigation technologies" are those technologies that reduce the levels of GHG emissions, covering short-, medium- or long-term periods. Many of the mitigation technologies focus on reducing carbon dioxide emissions from fossil fuel combustion as they form the largest share of overall GHG emissions. Commercially available technologies can play a large role in meeting the short-term goals of the Kyoto Protocol, as well as the goal of the UNFCCC over the medium term. The wedges focus on technologies that have the potential to produce a material difference by 2054, and include efforts beyond business-as-usual in the areas of carbon dioxide capture and storage (CCS); renewable energy ; energy efficiency and conservation ; fuel switching; nuclear fission; forest management; and agricultural soils management. Many of these available technologies are not yet fully competitive under current policies and markets.

There is a significant potential for new investments in the short term to utilise more efficient demand-side technology as energy efficiency gains translate into absolute reductions as well as a reduction in the carbon dioxide intensity of economies. Advances in energy efficiency and renewable energy technologies are required to enhance energy security, an increasingly important issue. This is a critical issue for many least-developed nations that are dependent on imported oil for their energy needs, and access to energy is a key consideration for economic development and poverty alleviation in most developing countries.

Energy security is also important for many developed nations, especially those that are net importers of oil and gas and reliant on supplies from around the world. Disruptions in the energy supply could have serious impacts on the economy and security of a nation.

Finance portfolio theory is now being used to show that diversifying the electricity generation mix at the national level to include. Renewable energy may have a higher capital cost, but it reduces overall system risk and cost, as its costs move independently of fossil fuel prices.

The production of renewable energy sources is expected to rise, but these sources are projected to meet only 2.5 per cent of overall global demand in 2030 under a business-as-usual scenario (IEA 2004). The potential market for renewable energy is impressive in developing countries because of climate, untapped potential, geographic size and undeveloped power infrastructure. Advanced renewable energy technologies, including those in the areas of small hydro, biomass and photovoltaic technologies, will be in high demand in all developing countries. The DD&D of renewable energy technologies can have significant economic, social and health benefits.

There are socioeconomic and geographic considerations in regard to technology DD&D. For example, China and India have large coal reserves, and economic development and energy security priorities will ensure these two countries account for two-thirds of the increase in world coal demand over the next 30 years. Given the size and projected growth rates of these two economies, clean coal technologies and carbon capture and storage (CCS) will be absolutely critical for limiting global GHG emissions.

Nuclear technologies can offer near-zero emission options, and China, Japan, Korea and India all plan to build more capacity. Most developed nations have not invested heavily in nuclear technologies since the 1970s, and public acceptability may inhibit their uptake in the short term. Over the longer term, more advanced technologies will emerge that have the potential to further significantly reduce GHG emissions. Ultimately solving the climate problem will require the decarbonisation of the world's energy supply, and many contend that revolutionary energy technologies will be needed, indicating that R&D investments are required now to ensure the necessary long-term technology DD&D. For example, technologies such as CCS, fuel cells, hydrogen systems, other renewables, biotechnology and advanced vehicles have the potential to be transformational technologies in future global energy systems if cost-effective technologies are available that meet other societal requirements including health, safety, reliability, other environmental concerns and cost. There is no one clear-cut technology option for moving forward and many areas of overlap exist among technology categories, where the success of one technology depends on the success of others.

Adaptation Technologies

Technologies for adaptation are those that "reduce the vulnerability, or enhance the resilience, of a natural or human system to the impacts of climate change". Examples include agricultural and forestry practices, coastal zone management, watershed management, and disaster reduction and preparedness. This includes both "hard" and "soft" technologies. Technologies for adaptation will be required both in the short term as impacts of climate change are already being noted, and in the long term as the need to adapt will grow over the coming decades. Technologies for adaptation are particularly important for developing nations, where climate change will further pressure "survival" needs such as access to drinking water and food security, as well as pose a threat to environment and heath standards. The macroeconomic costs of climate change are uncertain, but have the potential to seriously threaten development in many countries.

Effective technology DD&D that leads to reduced GHG emissions and improved adaptation can lead to significant co-benefits and assist countries in meeting development goals. The health impacts of a changing climate are likely to be overwhelmingly negative, and are associated with extreme weather events, increased air pollution, and water- and food-borne enteric diseases. Technologies for adaptation can assist countries in coping with extreme weather events, and improved air quality, which results from decreased burning of fossil fuels, can reduce the cost of health care expenditures associated with air pollution-related illnesses. A report in the Lancet estimated that strategic climate policies could prevent about eight million deaths globally that would occur between 2000 and 2020 in a business-as-usual scenario.

Technology Development, Deployment and Diffusion

The salient point is that actions to date offer valuable lessons for the development of effective initiatives to enhance technology DD&D. Successful DD&D requires cooperation among stakeholders-including multilateral and bilateral development agencies, international and regional financial institutions, governments, research institutions and the private sector. The creation of an enabling environment can assist in bringing together partners and creating the conditions for climate-friendly investments and focused initiatives to link climate change mitigation and adaptation with broader sustainable development goals. A key challenge for the post-2012 regime will be to identify and implement technology DD&D actions that build on current initiatives to ensure that climate change is a key consideration in investment decisions to allow for significant technology change over the medium to long term. This will require significant mobilisation of public and private sector finance toward zero and low-carbon technologies, including a determined shift toward demand- and supply-side energy efficiency.

Barriers and Strategies

Barriers

A number of barriers must be overcome to enhance technology DD&D. The EGTT has undertaken extensive work in analysing barriers to technology transfer, including hosting workshops on technology needs assessments, enabling environments and innovative financing. The issue was explored in the 1998 UNFCCC technical paper, Barriers and opportunities related to the transfer of technology.

As well, a senior-level round table on enabling environments for technology transfer was held on December 8, 2003 in Milan, Italy at COP-9, where discussion highlighted, inter alia, different technology transfer experiences, including macro and micro barriers. "Barriers exist at every state of the technology transfer process-technical, economic, political, cultural, social, behavioural and institutional." Challenges to the effective deployment and dissemination of technology in developing countries include the need to adapt technologies to local conditions, lack of supply chains, insufficient human capabilities and lack of integral institutional structures, such as stable grids and stable investment climates.

The UNFCCC workshop on enabling environments identified commonly encountered barriers to technology transfer, including failures in "reflecting economic and environmental costs in prices; enforcing regulations; ensuring awareness of relevant measures; and developing affordable cleaner technology" as well as the "high cost of patented technology, limited short-term profitability of some environmentally-sound technologies, limited finance, insufficient technological know-how, and inadequate institutional capability." The report on the UNFCCC workshop on innovative financing outlined the main barriers to accessing finance for technology transfer and noted that "many investors view renewable energy and GHG projects as compounding risk-combining risky sectors with risky markets with a risky commodity," and that future work should focus on removing the barriers in developed as well as developing countries.

The work of the EGTT has tended to focus on barriers to technology transfer between developed and developing countries. Barriers can also be looked at in a wider context when assessing technology DD&D, which takes place within countries and regions, between developed countries, between developing countries, and between developed and developing countries. For example, technology lock-in is a major barrier to the rapid uptake of new climate-friendly technology in all of the above circumstances. Established systems have market advantages arising from existing infrastructure, services and institutions. Fundamental shifts in technology take time, and DD&D can take decades before new technologies become widely accepted and economically competitive.

In developed countries, social and behavioural preferences for existing technologies and lifestyles at the household level are a major barrier, often reinforced by media and advertising. Within the business and industry sector, some may resist politically-driven technological change if there are insufficient incentives. Once built, large units of physical capital can operate for many decades. In the near term, patterns of capital consumption are driven by factors largely unrelated to climate change, and will likely only use climate change as an investment decision factor when obligated to do by regulations.

New technologies often face a major disadvantage when competing on a direct-cost basis with conventional technologies; they generate fewer environmental externalities than conventional technologies, a benefit which is typically not reflected in market prices. Thus, mechanisms are required to provide the right signal about the full social costs of technology choices. Price signals that reflect the negative environmental and social externalities related to energy use can improve the DD&D of climate friendly technologies, as well as influence the diffusion of energy-using technologies that are being innovated and commercialised at a rapid rate.

Neuhoff argues that an uneven playing field exists because of direct and indirect subsidies. He estimates that OECD countries spent US$20-30 billion on energy subsidies in 2002 and such subsidies may delay investment in energy efficiency and renewable energy provision. Traditional export credit guarantees by OECD countries benefit traditional energy technologies. In the late 1990s, Export Crediting Agencies (ECAs) facilitated a US$17 billion annual investment in fossil energy and only US$0. 8 billion in renewables. Neuhoff notes the lock-in of technologies is exacerbated by the lower production costs for conventional technologies and the cost of adopting new, more efficient technologies.

Weak intellectual property rights (IPR) regimes are often perceived as barriers to technology DD&D, as technology owners fear their technology may be stolen if they sell equipment without a license agreement. Development of IPR regimes in developing countries can assist those countries in attracting private sector investment, as well as encouraging innovation and the development of indigenous technologies. Many developing countries face additional challenges when trying to attract investment into the energy and infrastructure technologies needed for social development due to the perception of high risk and inadequate returns, alongside substantial transaction costs.

Exchange rate risks are also a factor in projects where revenue is generated in local currency but fuel must be bought in foreign currency. The resulting higher risk premium means returns have to be "significantly higher" than in OECD countries. This highlights the importance, at the national level, of creating the right enabling environments-through political and regulatory reform, together with greater local capacity-to stimulate private capital to flow to those regions. Increases in the production of renewable energy sources can also mitigate the exchange rate risk by reducing the inputs of foreign fuel. Developed and developing countries also face challenges in developing and attracting technologies for adaptation.

Strategies

Much research has been undertaken on overcoming barriers and creating strategies to more broadly enhance technology DD&D. Technology "push" and "pull" options will be required to bring existing and emerging technologies into implementation. Many experts note that carbon prices needed to meet the Kyoto Protocol will likely not be high enough to create the "pull" required to develop more advanced technologies in the short term. Thus, incentives will be required to support technology DD&D over both the short and longer term.

Innovative financing options. Innovative options are also required to attract financing for technology DD&D, and were addressed at the EGTT workshop on financing. A key message was that risk sharing can increase financial investment and stimulate local private-sector participation. Governments have an essential role in risk management, including establishing priority investment areas, setting out framework conditions for transactions and shaping financial flows and providing a stable legislative and regulatory environment.

Technology needs assessments can assist in identifying the key stakeholders and capacity building for local investors can assist in project preparation and development to improve access to finance. The workshop also emphasised the importance of seed capital, as well as the need to develop a "common language" for ongoing dialogue between governments and the finance sector. An effective policy framework will be "loud, long and legal," to ensure the right signals attract capital; the rules and incentives are stable and sustained; and that the regulatory framework provides the basis for long-life capital-intensive investments.

Private sector investment. Encouraging private sector investment is critical for technology DD&D, as private markets have increasingly become the main avenue for the transfer of technology. Developing countries are expected to account for 70 per cent of the growth in global energy demand and two-thirds of the growth in GHG emissions. Foreign direct investment (FDI) is the largest component of external funding to developing countries, and is expected to be the major player in financing energy and transportation infrastructure in developed as well as developing countries.

As FDI accounted for 60-80 per cent of global financial flows in recent years, private sector investment should be viewed as one of the main vehicles for ensuring that technology DD&D contributes to short- and longer-term climate change mitigation and adaptation, despite the fact that private investment remains low in many developing countries.

In this context, there is a need to ensure that the financial markets and public finance institutions are mutually reinforcing efforts to more effectively support sustainable development if the world is to stabilise GHG emission reductions over the long term. Public-private partnerships are increasingly being used as an innovative means to promote private sector investment in climate-friendly technologies. The private sector needs to be better engaged to support technology DD&D. Private sector involvement can be encouraged through the development of incentives for companies. The establishment of business networks and transparent competitive solicitation processes can also promote private sector investment. Reducing the risks of investing in projects in developing countries is essential; requiring improved institutional and organisational structures, transparent legal and regulatory frameworks, and improved economic management.

The promotion of technology DD&D to developing countries will require new and additional financial resources that can benefit both the technology provider and the technology recipient. Cost-sharing may be one means to increase private sector investment. Establishing the right price signals that account for negative externalities for could also assist in opening up markets for climate-friendly technologies, which are often not the lowest-cost option.

Multilateral financial institutions and export crediting agencies. Multilateral financial institutions and Export Crediting Agencies (ECAs) leverage substantial private sector investments. ECAs invest or indirectly support development activities that are considered too risky for the private and/or public sector. Investments in climate-friendly technologies may be eligible for support from ECAs, and there is opportunity to encourage coherence of these activities with global climate change goals through selectivity in the projects they support, the use of common approaches for evaluating the environmental impacts of projects and the introduction of energy efficiency and carbon intensity standards.

The International Finance Corporation, which is the private sector arm of the World Bank, is increasingly supporting regional environmental business facilities, particularly small- and medium-sized businesses that are active in the areas of energy efficiency and cleaner production methods, and other carbon finance opportunities.

Actions under the UNFCCC. Technology transfer between developed and developing nations has been an important topic within international climate change negotiations and some developing nations refused to sign the UNFCCC and Agenda 21 until developed countries more clearly committed to supporting technology transfer. Thus, technology transfer has been an incentive for developing country participation in the UNFCCC, and increased access to technology will continue to play that role in a post-2012 world.

The EGTT has identified and is working to improve technology transfer, and developed countries have supported technology transfer activities through both multilateral and bilateral channels. GEF support has been provided for capacity-building, technology needs assessments and technology transfer, including projects activities in the areas of renewable energy, energy efficiency and energy conservation. At COP-10, developed countries pledged US$34.7 million to the SCCF to finance projects relating to adaptation; technology transfer and capacity building; energy, transport, industry, agriculture, forestry and waste management; and economic diversification.

Policies and programmes at the national level. The IPCC notes that a comprehensive set of actions is needed to promote technology innovation, development and deployment. Policies and programmes at the national level can enhance technology DD&D, and national governments can use fiscal measures; regulatory policies; and information, labelling, voluntary and other assistance programmes to enhance technology DD&D. Lempert et al. argue that government funded research plays an essential role in developing new technologies and advancing new technologies to market readiness through such policies as tax credits, appreciated deceleration of investments reducing GHG emissions and government procurement of low-emitting technologies.

Government policies can assist in reducing firms' uncertainty about new technologies and should be designed to enhance "learning by doing" to increase the likelihood that new technologies will be deployed during times of rapid capital turnover. Governments can also induce firms to retire old technologies through regulatory reform, infrastructure policies, and promoting information dissemination programmes that encourage environmentally-conscious consumer demand.

The OECD and IEA promote these and other policy tools, such as cooperating with the private sector to develop and diffuse new technologies, facilitating public-private and inter-firm collaboration, seeking out opportunities for international collaboration and providing access to "learning opportunities". To ensure the involvement of the private sector in technology DD&D in developing countries, activities such as fair trade policies, protecting IPR, diversifying forms of assurance mechanisms, reducing transaction costs of collaboration and increasing public access to information about technologies are necessary. Flannery noted key roles for the public and private sectors at the 2004 IPCC Expert Meeting on Industrial Technology Development, Transfer and Diffusion. He stressed the need for governments to establish enabling frameworks to allow the private sector to bear the risks and capture the rewards of deploying technologies.

Official development assistance. The issue of climate change would significantly benefit from a fuller integration into the mainstream activities of development agencies. While a number of focused activities have assisted countries with National Communications and National Adaptation strategies, linkages between climate change and poverty alleviation are still not fully appreciated. This is beginning to change and increasingly initiatives of development agencies support the mitigation of and adaptation to climate change. There is substantial room for ODA to contribute to technology DD&D through projects in the areas of energy infrastructure, energy efficiency, transportation infrastructure, agriculture and forestry.

ODA is particularly well positioned to assist in technology DD&D in areas where the private sector is not active, such as adaptation or in gaps in the financing supply chain that create blockages to the development of commercial operations. ODA can also be used to enhance enabling environments and provide capacity building to create the necessary foundation and absorptive capacity for successful technology DD&D.

Capacity building can also assist countries in identifying new technologies for their specific needs and the benefits associated with new technologies, such as improved productivity, cleaner air and water and related health benefits. For technology DD&D to be effective, it must address broader economic development and quality of life concerns. Many countries are facing increased problems with environmental degradation and clean technologies could help to improve health conditions.

Technology Development, Deployment and Diffusion in a Post-2012 Regime

A number of options for enhancing technology DD&D in

a post-2012 regime have been put forward. There is no one technology or action that will induce the required emission reductions to meet the goal of the UNFCCC, but rather a suite of actions is required-of which technology DD&D will be one component. It is important to note that the UNFCCC is only one of a number of actors promoting technology DD&D, which is a broad issue that encompasses not only climate change, but also economic development, poverty alleviation, sustainable development and energy security.

Philibert explains that actions to promote technology DD&D have the potential to ease barriers to strengthened emissions mitigation cooperation by promoting a deeper understanding of difficulties faced by countries, building confidence between countries, increasing the relationships between stakeholders in and between the various countries, and remaining engaged on common mitigation action even though countries may have difficulties agreeing on a global framework. Many experts agree that actions to promote technology DD&D will be most effective if they are a complement to a credible, global commitment to limit GHG concentrations. An exclusively technology-focused approach is unlikely to provide the emission reductions required, but should be considered as one component of a future climate regime.

Technology Agreements and Protocols

Technology agreements and protocols are one option for promoting the DD&D of climate-friendly technologies. International commitments may relate to the use of common technology standards, such as energy efficiency standards for appliances, the prescribed use of low or zero-carbon technologies or minimum standards for the energy efficiency of industrial processes or power plants. A number of proposals that could be developed under or outside the UNFCCC negotiating umbrella have been put forward regarding technology agreements and protocols.

Technology deployment and diffusion

Barrett argues that common technology standards will provide incentives for investments in climate technology, as well as provide incentives for compliance and participation. He suggests that technology standards will be largely self-enforcing because if enough countries adopt the standards, other countries will follow common standards because of economies of scale in production and network effects. The most attractive approach would be to establish technology protocols in such areas as the use of hybrid engines, fuel cells or standards for fossil fuel power plants to capture and store carbon. Barrett's proposal would have developing countries be bound by technology standards set out in separate protocols, but the diffusion of required technologies would be financed by developed countries, with contributions based on ability and willingness to pay as determined by the UN's scale of assessments.

Benedick proposes a portfolio approach that emphasizes long-term international standards and incentives for technology development and deployment that aims to promote a technological revolution in energy production and consumption. The approach includes the adoption of a portfolio of policies that would be coordinated with like-minded nations and include technology targets for power generation and fuel efficiency standards for automobiles. The portfolio would include a carbon tax to fund new technology research and a programme funded by developed countries to promote technology transfer to developing countries.

Edmonds and Edmonds and Wise put forward a proposal for a technology backstop protocol, that would serve as a backstop in the event that other options failed. A technology based protocol would set out medium- to long-term international technology targets and/or standards for new fossil-fuel electric power plants and synthetic fuel plants installed in industrialised countries after 2020. Developing countries would be required to do the same upon reaching identified levels of development, defined as when their per capita income has risen to the average 2020 income level for industrialised countries in purchasing power parity terms.

Tol proposes a technology protocol that would specify the speed at which Best Available Technologies (BAT) standards would progress, with standards only applying to developed countries. Ninomiya proposes an international agreement on energy efficiency levels in major emitting industries and energy efficiency standards for major appliances in the residential and transportation sectors. The proposed agreement would complement the Kyoto Protocol and aim at participation by developed and major developing countries. He suggests that a global R&D fund should be considered to support the development of appropriate technologies.

Technology R&D agreements

Many experts have indicated that increases in Research and Development are critical to develop new technologies for a carbon-constrained future. Technology R&D agreements can be useful components of a future climate regime by enhancing the long-term perspective and effectiveness, and making use of market forces. Cooperation on R&D can allow participants to benefit from each others' efforts, help disseminate technologies, and reduce the costs of such efforts by sharing results and preventing duplication of efforts.

Barrett has proposed an international agreement on R&D funding as the "push" component to his proposal for agreements on technology standards. He proposes common research efforts for the development of climate-friendly technologies, whereby countries make a financial commitment to research programmes based on ability and willingness to pay. Incentives for participation would include the sharing of R&D results among participating countries only. Barrett proposes a research emphasis on electric power and transportation. Sanden and Azar put forward a similar proposal whereby R&D is supported through an R&D carbon levy of US$0. per tonne of carbon to raise revenues.

Similar to technology agreements to promote deployment and diffusion, R&D agreements offer potential flexibility in the design of a post-2012 climate framework as such agreements could be undertaken under the UNFCCC or as a separate regime. The strengths of R&D agreements, as noted by den Elzen and Berk, include an enhanced long-term perspective and enhanced technological capacity in participating countries. The weaknesses include uncertainty on environmental effectiveness, lack of market incentives to apply technologies and the risk of selecting and promoting less effective technologies.

Regional Sectoral Agreements

National or regional sectoral agreements could be related to limitation or reduction commitments for levels of GHG emissions or energy use. Watson et al. note that sectoral agreements could be designed to encourage climate-friendly investments in certain sectors, or to encourage the adoption of climate-friendly policies in a sector through taxes, standards or other regulation. Schmidt et al. explain that sector-based approaches can be developed in a variety of forms, such as "fixed-based limits, dynamic targets, benchmarked-based, harmonised policies and measures or combinations." Sectoral agreements could be global, defining commitments or targets for specific sectors such as cement or steel production. These types of targets could work for internationally-oriented sectors with a fairly limited number of actors, and the commitments could consist of emission limitations or reduction targets for the entire sector, or process-related targets such as the use.of low-emission technologies.

Sectoral targets and commitments offer advantages when viewed in an international climate regime. Such agreements could be undertaken under the UNFCCC or as a separate regime, and could be voluntary or mandatory. Sectoral agreements may be easier to negotiate than other technology agreements, as they help to create a level playing field for international sectors, and could enhance technology spill-over and transfer. They are also suitable for developing countries as it allows them to address GHG emissions in a step-by-step manner. Weaknesses include the potential for carbon leakage to other sectors without targets if substitutes to the products are available, and standards run the risk of being either unambitious or unachievable because regulators do not know the exact amount of improvement that is feasible.

Bilateral Technology Initiatives

Agreements to cooperate on climate technology are taking place outside of the UNFCCC, and a number of countries have signed bilateral agreements on technology and scientific cooperation. Buchner and Carraro note that the European Union cooperates on international science policy with more than 30 countries and is engaged in a number of technology agreements aimed at the improvement of energy technologies and climate-friendly production processes; the United States has signed various bilateral climate technology agreements ; and Japan has strengthened its role in climate cooperation with Asia, including a joint research initiative with seven developing nations aimed at providing technological assistance to these countries to reduce their GHGs in exchange for carbon dioxide emission credits.

The Asia-Pacific Partnership on Clean Development and Climate aims to promote the development, diffusion, deployment and transfer of existing and emerging cost-effective cleaner technology and practices. This partnership is to be consistent with and contribute to efforts under the UNFCCC and will complement the Kyoto Protocol. Regional and bilateral initiatives can promote technology DD&D between developed countries, between developed and developing countries and between developing countries. Support for South-South cooperation can be a cost-effective means of assisting developing nations.

Technology Deployment Mechanisms under the Kyoto Protocol

Technology deployment mechanisms exist under the Kyoto Protocol, including the CDM. Suggested enhancements to the CDM to encourage technology diffusion include an expansion to include policy-based CDM, sectoral crediting mechanisms, and the Japanese proposal regarding credits for technology. Sectoral crediting mechanisms would provide incentives for developing countries to develop national, regional or sectoral projects, based on the adoption of policies and measures, rather than project-based investments, to achieve emission reductions. Examples include the modernisation of country's cement sector or reduction of emissions in the transportation sector. Policy-based CDM would grant credits to governments that enacted GHG-reducing policy reforms.

A government might, for example, adopt a particularly strong efficiency standard in its building code, saving on the energy used in heating and cooling. The resulting emission reduction would be credited to the government. Japan plans to present a proposal at COP-11/MOP-1 in which industrialised nations would be able to transfer energy-saving technologies to developing nations as part of emissions quota transactions.

Joint Implementation (JI) also offers opportunities to encourage technology deployment between developed nations and economies in transition. Many Eastern European nations and former Soviet states view JI as a means to attract investment from OECD nations, as well as fulfil international commitments under the UNFCCC as they anticipate that their own countries and firms will be able to use part of the reductions to help fulfil their commitments. Many of the economies in transition aim to increase energy efficiency and reduce the environmental impacts of energy production and consumption, as many of these countries have severe problems with acid rain, urban smog and the health impacts of air pollution. JI may be able to assist with technology DD&D by encouraging projects in areas that may be difficult to secure financing, and assisting in attracting private capital.

Framework for Technology Transfer

Technology DD&D could be encouraged through increased support for the framework for technology transfer established under the UNFCCC. This could assist in technology development and deployment in developed and developing countries by building on the work led by the EGTT in the areas of technology needs assessments, technology information, enabling environments, capacity building and mechanisms for technology transfer. Important work has been undertaken to identify technology needs and key stakeholders in the technology transfer process, develop a technology information system, identify barriers and develop strategies for governments to overcome barriers, and build capacity to identify appropriate technologies for local conditions. These actions lay the groundwork for broader technology DD&D, including attracting privatesector investment.

Even if an international technology cooperation agreement were adopted, a large technology transfer role would remain, both to fulfil developed country obligations under Article 4.5 and to provide advice and support to developing nations in need, particularly the least-developed nations which are not well-positioned to attract private investment. Actions to promote technology DD&D that involve developing countries, such as international technology cooperation agreements, should build on the work of the EGTT.

Technologies for Adaptation

While technologies for adaptation could be, and should be, included under broader technology agreements and in technology transfer efforts, specific actions have been recommended to support the DD&D of technologies for adaptation in a post-2012 regime.

— *Three Track Global Framework proposal*—includes an adaptation track that would build upon the Marrakesh Funds, be funded by industrialised countries and include compensation for damages;

— *Global Climate Agreement*—includes funding provided by Annex I Parties for building the adaptive capacity of vulnerable countries in line with the "polluter-pay" principle of the UNFCCC; support for capacity building in developing countries in areas such as development of sector-specific adaptation strategies;

modification of GEF rules to better facilitate adaptation projects that provide local benefits and increase capacity to access to funding; and innovative insurance schemes; and

— *Broadening the Climate Regime*—puts forward an Adaptation Protocol designed to secure the transfer of funds and technology to those countries most vulnerable to the impacts of climate change.

Menu of Options

The preceding discussion put forward a menu of options related to how the global climate regime can more strongly promote technology DD&D. Technology cooperation to support DD&D will be a long-term approach, but in the short term it may serve as a complement to other efforts to reduce emissions and adapt to the impacts of climate change. To create the required long-term effects, it will be necessary to start directing investment toward climate-friendly technologies in the short term and to persist with these investments thereafter. Moving forward in this regard could include a combination of approaches, as well as identifying options to build on work completed to date by building synergies with existing programmes and institutions.

Technology agreements will be a critical component of any future climate regime if it is to be effective, irrespective of whether or not such agreements are better negotiated within or outside the formal parameters of the UNFCCC. Clearly, outside efforts are playing an increasingly important role: witness the many ongoing efforts, including: the G8 Action Plan on Climate Change; Clean Energy and Sustainable Development; the Asia-Pacific Partnership on Clean Development and Climate; implementing agreements of the IEA; and the Methane to Markets partnership. There is opportunity for technology agreements among regional or likeminded partners to create niche markets, which in turn can bring costs down and assist in diffusion to the rest of the world. A global framework under the UNFCCC that recognises a variety of technology agreements could assist in legitimizing their activities and ensure recognition of DD&D actions under the agreements.

The most likely approach for the post-2012 regime to encourage technology DD&D would be the development of a technology agreement that includes developed countries in the short term, with economies in transition and developing countries having to satisfy the same obligations when they meet a certain level of development. A second viable option for the post-2012 regime would be an international R&D technology agreement, whereby developed nations commit to increase R&D spending and work in partnership with developing countries. Sectoral, technology cooperation and R&D agreements could be negotiated as an amendment to the UNFCCC, or alternatively be negotiated beyond the UNFCCC between like-minded countries as a demonstration of commitment to action to increase technology DD&D.

Technology DD&D is inherently a long-term approach, yet there are actions that can be undertaken in the short term. In particular, a viable alternative might be to seek agreement at COP-11/MOP-1 on the need to begin discussions on an agreement to promote technology DD&D and how such actions can be meaningfully reflected as part of the broader commitments in the post-2012 regime.

13

International Effort aganist Climate Change

Importance of Commitments

The nature of the climate change problem, as well as the history of international environmental cooperation more generally, suggest the need for commitments. The existence and implications of purported legal obligations, such as the duty to prevent transboundary pollution and the polluter pays principle, are the subject of endless debate among scholars and states.

Although these principles reflect strong moral imperatives-and may even have the status of international law-in the absence of courts that could apply and enforce them, they are unlikely to be of significant use in changing states' behaviour. Instead, states are likely to address climate change only if they believe it is in their interest to do so. That is why climate change negotiations have focused on "commitments," requirements that a state itself assumes, rather than on "obligations," a broader term that includes norms externally imposed.

The role of commitments derives from the "collective action" nature of the climate change problem. Like other collective action problems, climate change mitigation poses a fundamental dilemma. Because most of the benefits of climate change mitigation do not accrue to the country taking action, but are instead shared by the international community as a whole, individual countries have little incentive to do anything on their own. Even when the global benefits justify the costs, the country engaging in mitigation usually receives only a fraction of the total benefits. So, from its individual perspective, the costs of mitigation are likely to exceed the benefits.

Of course, it the costs of reducing emissions are sufficiently low, countries might be willing to go ahead anyway, for example, to show leadership or for public relations purposes. But significant investments to reduce greenhouse gas (GHG) emissions will be in a country's individual self-interest only if they are reciprocated by other states-only if a country's actions are part of a bargain involving significant action by others to address climate change.

International commitments serve as the glue that helps hold a cooperative regime together. Before taking potentially costly actions to address climate change, states need to be confident that others will do their part as well. International commitments are the means by which countries bind themselves to one another to take mutual action. What does it mean to say that a country "commits" itself to undertake mitigation actions?

In one sense, virtually all international commitments are voluntary. Given the absence of an international legislature that can impose obligations on states, international obligations in general depend on a state's consent. But, by making a commitment, a state agrees to limit its future freedom of action; it promises to behave in a certain way or to achieve a certain result. While its acceptance of a commitment is voluntary, its fulfilment of the commitment is not.

International commitments fall along a spectrum. Some are political, such as the aim in the UN Framework Convention on Climate Change (UNFCCC) to return developed country emissions to 1990 levels by the year 2000; others are legal, such as the reporting requirements in the UNFCCC and the targets and timetables in the Kyoto Protocol and the Montreal Ozone Protocol. In the absence of effective institutions to interpret and enforce international law, the distinction between political and legal commitments can often seem illusory.

Most international commitments-even "legally-binding" ones-depend on the good faith of states and on the diffuse costs of developing a reputation for breaking one's promises, which makes it more difficult to enter into mutually-advantageous deals in the future. But, in general, casting a commitment in "legal" form signals a greater level of seriousness by states, raises the costs of violation, and sets in motion domestic legal implementation mechanisms.

Of course, no level of commitment can fully assure that a country will uphold its end of the bargain. Some countries may view their treaty commitments as aspirational rather than absolutely binding. But, compared to a strictly voluntary system, commitments provide states with greater confidence that other states will not simply say one thing and then do another. This not only promotes action by states, but provides a signal to the market that helps drive changes in private behaviour.

Moreover, if mechanisms can be agreed to impose specific sanctions for violations, this further raises the costs of noncompliance and thus provides additional assurance to states that others will comply with their commitments. Indeed, given the potentially high short-term costs of mitigating climate change, many analysts believe that both legally binding commitments, and a strong compliance system are essential.

VARIABLES

All three of these variables-who, what, and when-raise important, interdependent issues.

Types of Commitments

Specifying the content of a commitment has both formal and substantive dimensions:

Binding vs. Non-Binding commitments

To begin with, there is the issue of the legal form of a commitment-in particular, whether it will be legally binding or political. This is not simply an either-or choice; a range of options present themselves: Although perhaps strictly speaking a misnomer, a "commitment" can be expressed in non-legally binding language, as a recommendation or an aim. The emissions target for developed countries in the UNFCCC was contained in the commitments section of the treaty, but was stated as an "aim" rather than a legal requirement.

One-way commitments

This is a variant of the previous option. An aim, although nonbinding, could have legal consequences in the sense that, if bettered, it can provide a country with certain legal benefits. Project baselines established under Kyoto's Clean Development Mechanism (CDM) are, in essence, one-way "commitments," since a country faces no penalty if its project exceeds a baseline, but receives certified emission reduction credits if the project reduces emissions below the baseline.

Legally binding commitments

A commitment can also be expressed in binding language ("shall"), like the targets and timetables in the Kyoto Protocol. It is important to note that this is a separate question from whether the commitment is subject to enforcement through a compliance system. Most international commitments do not have any specific compliance mechanisms. Nonetheless, they are legally binding and must be complied with by those states that accept the commitment.

Enforceable commitments

A binding commitment can be subject to a mandatory compliance system, with authority to respond to violations, such as the dispute settlement system adopted under the World Trade Organisation. This would provide the greatest assurance of compliance but would also present the greatest worry for states that are on the fence about whether to undertake mitigation commitments.

Policy Instrument

The substantive content of commitments can involve an equally wide variety of policy instruments:

Emission targets

An emission target is an obligation of result: it requires regulated entities to achieve a particular level or rate of emissions, but allows them flexibility as to how they will achieve that result. Emissions targets can be specified in various ways: fixed or indexed, absolute or conditional, and economy-wide or sectorial.

— *Absolute targets* :Until recently, most of the attention in the climate change regime has focused on fixed, country wide emissions targets, pegged to an historical base-year emissions level. The Kyoto Protocol, for example, requires industrialised countries to achieve predetermined, fixed levels of emissions for the 2008-2012 commitment period. In this respect, the climate change regime has followed the approach used in several other international environmental regimes, including those addressing acid rain and stratospheric ozone depletion.

— *Indexed targets* :Because emissions depend on a wide range of variables that are difficult to anticipate in advance, an emission target can be pegged to one or more of these variables, rather than defined in fixed terms, like the Kyoto targets. Thus far, most of the literature has focused on tying emissions targets to a country's GDP so that the permitted level of emissions would be larger or smaller, depending on whether the economy grows or shrinks. The Bush Administration's carbon intensity target and the proposed Argentine target are both examples of indexed GDP-based targets.

— *Conditional targets* :In contrast to the Kyoto targets, which apply come what may, a target could be formulated in conditional terms: if the specified conditions are not satisfied, then the target either would not apply at all or would be modified in some fashion. One option is to make commitments conditional on a state's achievement of a minimum level of wealth. In addition, conditional targets-like indexed targets-could help alleviate fears that a fixed emission target might become an economic straitjacket. A conditional target that has received particular attention in this regard is the so-called "safety valve" approach.

— *Sectorial targets* :A target can also be specified on a narrower basis than total national emissions. For example, targets could be specified for particular sectors or industries that are particularly important, politically easier to address, or comparatively insulated from international competition. Sectorial targets could be binding or "no lose," fixed or indexed.

Financial targets

Rather than focus on emissions, a target can be specified in financial terms, as an amount to be devoted to climate change mitigation, either domestically or internationally.

Policies and measures

In contrast to a target-based approach, a commitment regarding policies and measures (PAMs) is an obligation of conduct rather than an obligation of result: it requires countries to act in certain ways, but does not require them to achieve any particular level of emissions or financial contribution. During the negotiation of the Kyoto Protocol, the European Union pushed for the inclusion of commitments related to policies and measures, but due to strong resistance from the United States, the Protocol includes only an illustrative list of possible PAMs, without requiring states to adopt them.

— *Technology and performance standards* :An international commitment can address the use of emission-reduction technologies. For example, it could specify mandatory standards relating to appliance efficiency, residential insulation, or the use of renewable or other non-emitting energy sources. The international commitment can either require the use of particular technologies or set forth a performance standard that allows private entities flexibility as to the choice of particular technologies.

— *Taxes* :An international commitment can provide for a common or harmonised tax on GHG emissions. So long as a country had the required tax in place, it would satisfy its international commitment, regardless of the actual level of emissions reduction achieved.

— *Subsidy removal* :An international commitment can require countries to remove specified subsidies, for example, on energy production or consumption. The Kyoto Protocol includes in its illustrative list of PAMs for developed countries "the progressive reduction and phasing out of subsidies." Subsidies are a problem not only in industrialised countries: the International Energy Agency estimates that removing energy subsidies in just eight developing and transition countries would reduce their CO_2 emissions by 17 percent and global emissions by 4.6 percent.

— *Emissions trading* :An emissions commitment can be coupled with a PAM requiring countries to implement a domestic emissions trading programme with specified features. The European Union directive on emissions trading represents an effort of this kind: it sets forth the parameters of a required emissions trading system for EU member states.

— *Technology R & D and incentives* :To address the low rates of investment in research and development concerning emission-reducing technologies, a commitment might require states to devote additional resources for R & D, as well as for deployment of existing and new technologies. The agreement on the international space station is one illustration of an international agreement focusing on cooperative research, development, and deployment.

Since a targets-based approach and a PAM-based approach are often seen as competitors, it is worth emphasising that they could complement one another: a target could be used to specify the overall result to be achieved, while PAMs could specify the means for reaching that result. Indeed, in some cases the relationship could be even stronger.

Timing of Commitments

Another critical question is the timing of commitments.

The international negotiations thus far envision a dynamic process beginning with the relatively modest but important reporting requirements in the UNFCCC, to be followed by specific mitigation commitments in subsequent protocols. A future agreement could set forth a more detailed road map for the evolution of commitments over time. There are two important elements to timing: first, when will a commitment take effect, and second.

Initial stage of commitment

In contrast to most treaties, which set forth commitments that take effect immediately upon the agreement's entry into force, the Kyoto Protocol establishes a commitment period beginning more than ten years after its adoption. The intent was to avoid economic disruption by giving countries and firms time to adjust to the Kyoto targets. Even so, many economists argue that, if the United States had stayed in the Kyoto system, the Kyoto targets would have cost more than necessary by requiring premature capital retirement.

According to this view, an even longer-term target, timed to coincide with ordinary patterns of capital turnover, would have been more economically efficient. If a commitment is too far off in the future, however, it may lack credibility; it may raise concerns that, given the lack of stability in international politics, the commitment is likely to be changed before it ever takes effect.

Duration of commitment

In most international environmental regimes, commitments have an indefinite duration; they continue in effect until the parties modify or terminate them. The Kyoto Protocol, in contrast, defines an emission target for only a five-year period, ending in 2012. This is sometimes justified as providing necessary flexibility. The rationale is that, given the significant uncertainties relating to climate change, the international regime should consist of a series of rolling commitment periods, which allow commitments to be continually redefined to take account of improved scientific and economic understanding.

Most international environmental agreements have flexible amendment procedures, so that commitments can be periodically updated in response to new problems and new information. Similarly, the international trade rules and tariff rates set forth in the GATT/WTO regime are not time-limited. But this has not meant that they are carved in stone; instead, the trade regime has undergone major changes through periodic negotiating rounds.

The real effect of making commitments with a limited duration is to reverse the ordinary presumption of continuity. In other regimes, commitments continue until they are changed; in the Kyoto Protocol, they lapse unless they are renewed.

Subject to Commitments

Individuals entities

Although the climate change regime has, thus far, sought to establish obligations only for states-for example, relating to emissions targets, financial contributions, and reporting-an international commitment could conceivably apply directly to individuals, private entities, or sub-national entities such as cities. International criminal law, for example, establishes basic duties on individuals, the violation of which results in international criminal liability. Although individual criminal responsibility seems clearly inappropriate for climate-related activities, other forms of individual liability are possible.

It should be emphasized, however, that attempting to impose obligations directly on individuals or private entities would pose very difficult issues of implementation and enforcement-particularly with respect to individuals and firms located in countries that do not participate in the international regime and that therefore could not be enlisted for enforcement purposes. There are, at present, no examples of international environmental regimes that apply directly to individuals.

Commitments for states

Given the difficulties of imposing obligations directly on individuals, most international regimes define commitments for states and rely on them to translate these into obligations for individuals and firms under their jurisdiction. Because of the global nature of the climate change problem, the natural tendency is to include all countries in an international climate change regime.

All countries have a duty to participate because of their contribution to climate change, and they all have a right to participate because they will all be affected by it.

The UNFCCC takes this approach: it is open to any state and defines at least minimal obligations for all participants. At the same time, it recognises that the same level of commitment is not appropriate for all states. It therefore sets forth differentiated obligations, based on the principle of common but differentiated responsibilities and respective capabilities.

In establishing new commitments, a key question will be whether they apply equally to all states, or whether differentiation is appropriate. Kyoto's mitigation commitments all take the same form, for instance, but apply only to developed countries and vary in stringency among them. Commitments could also be differentiated by form ; by time frame ; or by conditionality.

The criteria that might be used to determine who should participate in a climate regime, or to differentiate commitments among the participants, include the following:

— *Big current emitters* :Relatively few countries contribute significantly to climate change- 15 countries, for example, account for 75 percent of global CO_2 emissions. Mitigation commitments by these big emitters could largely address the climate change problem. Moreover, limiting membership in the regime to countries with mitigation commitments could simplify the negotiating dynamic significantly.

— *Big historical emitters* : Alternatively, commitments might vary depending on a country's historical contribution to the climate change problem. Here, the rationale for differentiation would be the idea that countries with high historical emissions are responsible for the current problem and have a duty to fix it-including through reductions in their current emissions. This is the essence of the so-called "Brazilian proposal" for allocating the burdens of addressing climate change.

— *Rich countries* :Commitments could vary depending on a country's wealth and therefore its capacity to respond to the climate change problem.

— *Like-minded states* :A future climate regime could be limited to like-minded states, which are willing to undertake a certain level of commitments and have shared views about international implementation mechanisms such as emissions trading. Again, the idea would be to create a more favourable negotiating dynamic by conducting negotiations initially among countries with shared goals, bringing other countries in later.

Different Assessment Criteria

Potential commitments need to be evaluated from both a policy and a political perspective. In some cases, synergies may exist between different assessment criteria: a climate policy that is equitable or cost-effective may in the long run be more environmentally effective. But, often, different assessment criteria will be in tension.

Ensuring predictability in the costs of mitigation measures, for example, comes at the expense of predictability concerning environmental effects. More broadly, there are strong tensions between the basic goals of policy optimization and political feasibility. Formulating a sound climate change policy is not so difficult; nor is formulating a politically acceptable one. The challenge is to devise a policy that is both sound and acceptable.

Policy Assessment Criteria

There are five key criteria: environmental effectiveness, cost-effectiveness, equity, dynamic flexibility, and complementarity.

Role of environmental effectiveness

Ultimately, the purpose of mitigation commitments is to reduce dangerous climate change. The bottom-line test of commitments is their effectiveness, over the long run, in preventing climate change. An important contributor to environmental effectiveness is, of course, stringency-all other things being equal, a stronger commitment should produce a greater environmental result than a weaker one. But all other things are rarely equal and, as a result, environmental effectiveness is not solely a function of stringency. Other important factors include:

— *Leakage* :To the extent that the climate change regime is not global, private entities can avoid the impacts of commitments by shifting their operations to a nonparty state. As a result, more stringent targets could actually be counterproductive, both by discouraging countries from joining and by causing emitting activities to shift to states without commitments.

— *Stimulating technological change* :Some types of commitments may be more effective, over the long run, in inducing technological change. For example, many policy analysts argue that market based approaches, such as "cap-and-trade" or taxes, are more effective in promoting ongoing technological change than technology standards, which lock in a particular technology and fail to provide incentives for further change.

— *Changes in public attitudes, awareness, and learning* :Over the long run, addressing clim-ate change will likely require changes in public attitudes and behaviours. To the extent a commitment can help do so-for example, by raising public awareness-this would be an extra benefit.

— *Enforceability* Given the nature of the climate change problem, countries will be tempted to violate their commitments, since the near-term economic benefits of violation will typically outweigh the near-term environmental costs. For this reason, climate change commitments may be effective in changing behaviour only if they can be adequately monitored and enforced.

Role of cost-effectiveness

Since countries have only a finite level of resources to devote to climate change and other competing needs, commitments need to get the most "bang for their buck"; they need to reduce each unit of emissions at the lowest possible cost. Most economists agree that market based approaches-such as emissions trading and taxes-are best from this perspective. The more flexibility market participants have to seek out and utilise low-cost reduction options, the greater the economic effectiveness. That is why the Kyoto Protocol provides not only "where" flexibility, but also "what" flexibility and "when" flexibility.

Role of equity

Commitments should treat participants fairly. This is important not only in determining which commitments are politically acceptable; it is also an important end in itself. Whereas environmental and economic effectiveness can both be judged in absolute terms, equity is by its nature relational.

Role of Dynamic Flexibility

Given the likelihood that commitments will periodically need to be revised in light of new scientific and economic information, a commitment would ideally be formulated in a manner that allows revisions as needed.

Role of complementarity

The withdrawal of the United States from the Kyoto Protocol opens up the possibility of a fragmented climate regime, with different country groupings adopting different types of commitments. In that case, an important factor in assessing possible commitments would be the feedbacks, complementarities, and potential linkages between commitments in different regimes.

Political Assessment Criteria

From a political perspective, there are two key criteria: whether a particular type of commitment can be negotiated, and whether it can be implemented.

Commitments can be negotiated

In considering future commitments, the question is not simply which commitments are optimal, but which are negotiable. Most of the options for mitigation commitments discussed above have been proposed at one time or another. But none has been able to command a stable consensus. In some cases, an option may not be negotiable due to domestic political factors in particular countries.

But several more general considerations also affect the negotiability of mitigation commitments, including the following:

— *Continuity with Kyoto* :A commitment's continuity with Kyoto could cut both ways in terms of political acceptability. On the one hand, most countries now have a substantial investment in the Kyoto process, so a commitment's continuity with that process would be a point in its favour. At the same, Kyoto has become a negative icon for many in the United States, and is likely to remain a nonstarter even once a new administration takes office. In terms of this particular criterion, indexed or conditional targets could conceivably square the circle: they are compatible with the architecture established by Kyoto, including the emissions trading mechanism; but they are more flexible than the fixed, absolute targets in Kyoto, and thus could credibly be characterised as a different approach from Kyoto.

— *Economic predictability* :For countries as widely different as the United States and China, a primary concern with Kyoto-style commitments has been the possibility of high compliance costs. Although some economists estimate that the costs of compliance would be low- and that an emissions target for China could even be economically advantageous, given its potential to reduce emissions cheaply and to sell surplus credits to countries with higher mitigation costs- compliance costs depend on many unpredictable variables such as rates of economic and population growth and of technological change, which make economic estimates highly uncertain.

From a political standpoint, economic predictability may be as or more important than economic efficiency. Countries want to know in advance what they are undertaking and whether it makes political and economic sense.

— *Compatibility with sustainable development priorities* :Most developing countries perceive climate change mitigation and economic development to be in competition with one another: money invested in mitigation is money diverted from economic development. In the long run, developing countries will undertake climate change mitigation only if they see synergies with sustainable development goals.

Commitments can be implemented

To be effective over the long run, commitments need to take into account the capabilities and limitations of the institutions on which implementation and compliance will depend. The importance of institutional capacity is by now well understood in the context of technology transfer: the "best" available technology is not necessarily best for a country lacking the capacity required to use the technology effectively.

Instead, technologies that better fit a country's capacities may be more appropriate. At the international level, where institutions are notoriously weak, the issues of implementation and enforcement deserve particular attention.

Factors relevant to implementation include the following:

— *Ease of monitoring* : Different types of commitments vary widely in terms of the ease with which they can be monitored and verified. Some analysts attribute the success of the international oil pollution regime to its reliance on construction and design commitments that are easy to verify rather than on discharge standards. In the climate change context, national emissions of carbon dioxide can be estimated with a high degree of confidence, but emissions of other gases and removals by sinks are considerably more uncertain. Indexed targets introduce additional complexities, since they require monitoring not only of emission levels but also the variable to which emission allowances are pegged.

— *Predictability of compliance* : Most implementation of international commitments takes place at the national level, through national law, so commitments adopted internationally need to be capable of domestic legal application. One criticism of obligations of result, such as targets and timetables, is that, because compliance depends on changes in behaviour by firms and individuals, it is difficult for a country to predict accurately whether it will achieve the required result.

Future Commitments

Three caveats are in order. First, these options are, of course, not the only possibilities. Instead, they represent a range of approaches chosen to illustrate many of the general issues regarding mitigation commitments. Second, the assessments of the various options identify the most prominent advantages and disadvantages of each approach, rather than applying the assessment criteria discussed above in a systematic manner. Finally, these options could be combined in various ways; they are not mutually exclusive.

Sequentially-Negotiated National Emissions Targets

The Kyoto Protocol sets forth fixed national emission targets for the 2008-2012 period. The idea is that the first five-year commitment period will be followed by other commitment periods, to be negotiated on a rolling basis. Kyoto-style targets, if applied to all significant emitters, would have several benefits:

— *Environmental effectiveness* :Fixed targets, if complied with, provide the greatest environmental certainty.

— *Cost-effectiveness* : Fixed targets can be cost-effective if combined with emissions trading and with "when flexibility".

— *Equity Fixed* :targets can be differentiated among countries to meet equity concerns.

— *Dynamic flexibility/scalability Fixed* :targets can be adjusted up or down to take account of new information.

— *Continuity with Kyoto For* : countries that support Kyoto, fixed targets would provide the greatest continuity.

At the same time, absolute targets also have several significant drawbacks:

— *Difficulties of negotiating* :The costs of achieving a fixed national emissions target are uncertain, and depend on many factors that are difficult to predict. Although absolute targets can allow considerable flexibility in implementation, they represent a legal straitjacket in the sense that, once agreed, they do not provide for changing circumstances. This rigidity could make iterative negotiation of fixed short-term targets difficult.

— *Perceived incompatibility with development priorities* : Absolute targets are particularly problematic for developing countries and countries with rapidly growing economies, since they are seen as representing a potential constraint on economic growth. Of course, targets could build in "headroom" to allow developing country emissions to grow.

Targets for Indexed

Indexed targets have some of the same advantages and disadvantages as fixed targets. On the positive side, they are cost-effective if coupled with trading, which appears difficult but not impossible; they can be differentiated between countries and made more or less stringent as the circumstances warrant; and they could provide continuity with Kyoto.

Targets for Sectoral

Sectoral targets have the benefit over economy-wide targets of allowing states to proceed incrementally. Rather than attempt to develop a target that makes sense for the entire economy, states can address emissions in a step-by-step manner, starting with a more limited set of activities in sectors such as energy or transportation. That is why many national strategies for addressing GHG emissions take a sectoral approach.

Moreover, in some cases, more is known about emissions in one sector than another, so sectoral targets may help ease monitoring concerns. Finally, sectoral targets would make it more difficult for countries to give preferential treatment to particular sectors and, in that respect, could help ease competitiveness concerns.

But addressing emissions on a sectoral basis comes at a price. If states are restricted as to which types of emission reductions "count" internationally, they may be unable to take advantage of the most cost-effective options. Even if targets are developed for all sectors with significant GHG emissions, separate sectoral targets prevent countries and firms from making trade-offs across sectors, doing more in a sector where emissions can be reduced more cheaply and less in another sector where reductions are more expensive.

Targets for Hybrid

Hybrid targets, advocated primarily by economists, were put on the table informally by Brazil in 2000, during the negotiations that culminated in the Bonn/Marrakech Accords. Hybrid targets have a number of desirable features:

— *Economic predictability and negotiability* :By ensuring that the costs of mitigation commitments cannot rise above a predetermined level, hybrid targets remove one of the principal obstacles to the negotiation and acceptance of emission reduction targets.

— *Equity* :Although the safety valve level would need to be the same globally, commitments could still be differentiated through the emission reductions targets. Thus, the safety valve, like fixed targets, is compatible with the application of equity criteria.

— *Scalability* :A hybrid target could be scalable through its safety valve price as well as its emission reduction targets. To facilitate planning by business, the safety valve price could have an automatic escalator, which would apply unless the parties decided otherwise.

Of course, the economic predictability of hybrid targets comes at the expense of environmental predictability-the principal strength of fixed emission reduction targets. This has an obvious down side: if mitigation costs prove high and the safety valve kicks in, then the level of actual emission reductions would be less than under a fixed target. But there are risks either way. Just as we have no assurance what level of reductions a given price will buy, we have no assurance how much a particular emissions reduction will cost.

The difference is, the economic risks of excessive costs are near-term, while the environmental risks of insufficient reductions in emissions are longer-term and may be correctable through stronger measures later.

Targets For Developing Countries With Graduation Criteria

No-lose targets have been proposed primarily as a means of providing incentives for developing countries to accept emission targets. Over the long run, developing countries may need to accept binding targets as their economies develop. No-lose targets could serve as a useful transitional device, possibly in conjunction with criteria that define when a developing country would graduate from a nonbinding to a binding target. During the transitional period, no-lose targets could be combined with legally binding commitments in various ways.

Targets for Technology Standards

The difficulties involved in negotiating, monitoring, and enforcing emission targets have made technology standards more attractive, even to some economists who, as a rule, criticise such standards as inefficient. Technology standards-for example, relating to energy efficiency-could be negotiated by governments or through public-private partnerships.

One advantage is that they could have a significant environmental impact, even in the absence of universal acceptance, through tipping effects. As Scott Barrett explains: "If enough countries adopt a standard, it may become irresistible for others to follow, whether because of network effects, cost considerations... or lock-in." If so, technology standards would be essentially self-enforcing, and would not involve the compliance issues raised by emission targets.

At the same time, technology standards have a number of significant drawbacks that have limited their appeal in the climate change negotiations thus far. They depend on governments being able and willing to pick technologies based on sound technical considerations.

They lock in technologies and do not provide an incentive for further innovation. They limit flexibility by prescribing not just a result, but how countries must achieve it. For these reasons, among others, over the last decade, environmental policy has tended to move away from command-and-control regulation towards market based approaches.

Research and Development Commitments

If emission reduction technologies such as hydrogen fuel cells or carbon capture and storage became practicable and economic, this could go a long way towards overcoming the existing barriers to climate change mitigation. But recent studies indicate that, despite the high profile of the climate change issue, investments are going down overall in mitigation-related research and development.

International commitments by states to provide funding for research and development are not unprecedented. For example, the international space station is the product of an agreement providing for multilateral cooperation and funding. Voluntary approaches have also sometimes proven successful.

Twenty-one countries including the United States currently contribute to the Consultative Group on International Agricultural Research, which funds research centres around the world. So, while some countries such as the United States may be wary of any new financial obligations, financing of R & D might prove attractive, either as an alternative to more stringent types of mitigation commitments or, at a minimum, as an addon.

Bibliography

Arrhenius, Gustaf (1997). "Carbon Dioxide Warming of the Early Earth." *Ambio* 26: 12-16.

Baer, F., et al., Eds. (1991). *Climate in Human Perspective: A Tribute to Helmut F. Landsberg*. Dordrecht: Kluwer Academic.

Bolin, Bert, et al., Eds. (1986). *The Greenhouse Effect, Climatic Change, and Ecosystems. SCOPE Report No. 29*. Chichester: John Wiley.

Broecker, Wallace S. (2004). "Future Global Warming Scenarios." (Letter) *Science* 304: 388.

Crary, A. P., et al. (1955). "Evidences of Climate Change from Ice Island Studies." *Science* 122: 1171-73.

Crawford, Elisabeth (1996). *Arrhenius: From Ionic Theory to the Greenhouse Effect*. Canton, MA: Watson Publishing - Science History.

David, E.E., Jr. (1984). "Inventing the Future: Energy and the CO2 'Greenhouse' Effect." In *Climate Processes and Climate Sensitivity. (Geophysical Monograph 29, Maurice Ewing Vol. 5)*, edited by James E. Hansen and Taro Takahashi, pp. 1-5. Washington, DC: American Geophysical Union.

Emanuel, William R., et al. (1985). "Climatic Change and the Broad-Scale Distribution of Terrestrial Ecosystem Complexes." *Climatic Change* 7: 29-43.

Fonselius, Stig, et al. (1956). "Carbon Dioxide Variations in the Atmosphere." *Tellus* 8: 176-83.

Guilderson, Thomas P., et al. (1994). "Tropical Temperature Variations since 20,000 Years Ago: Modulating Interhemispheric Climate Change." *Science* 263: 663-65

Haigh, Joanna D. (1996). "The Impact of Solar Variability on Climate." *Science* 272: 981-84.

Handel, Mark David, and James S. Risbey (1992). "An Annotated Bibliography on the Greenhouse Effect and Climate Change." *Climatic Change* 21: 97-255.

IPCC (Intergovernmental Panel on Climate Change) (1996). *Climate Change 1995: The Science of Climate Change,* Edited by J.T. Houghton et al. Cambridge: Cambridge University Press.

Junge, Christian E. (1958). "Atmospheric Chemistry." *Advances in Geophysics* 4: 1-106.

Kalnay, Eugenia, and Ming Cai (2003). "Impact of Urbanization and Land-Use Change on Climate." *Nature* 423: 528-31.

MacCracken, M.C., and F.M. Luther (1985). "Detecting the Climatic Effects of Increasing Carbon Dioxide". Washington, DC, Department of Energy. Report DOE/ER-0235

Perry, John S. (1975). "The Global Atmospheric Research Program." *Reviews of Geophysics and Space Physics* 13: 661-67.

Rahmstorf, Stefan (1999). "Shifting Seas in the Greenhouse?" *Nature* 399: 523-24.

Schimel, D.S., et al. (2001). "Recent Patterns and Mechanisms of Carbon Exchange by Terrestrial Ecosystems." *Nature* 414: 169-72.

Thomas, Robert H., et al. (1979). "Effect of Climatic Warming on the West Antarctic Ice Sheet." *Nature* 277: 355-58.

Vellinga, M., and R.A. Wood (2002). "Global Climatic Impacts of a Collapse of the Atlantic Thermohaline Circulation." *Climactic Change* 54(251-67).

Wigley, Tom M.L., et al. (1986). "Warm World Scenarios and the Detection of Climatic Change Induced by Radiatively Active Gases." In *The Greenhouse Effect, Climatic Change, and Ecosystems. SCOPE Report No. 29,* edited by Bert Bolin et al., pp. 271-322. Chichester: John Wiley.

Young, William R. (2000). "The Future of Physical Oceanography." In *Fifty Years of Ocean Discovery. National Science Foundation 1950-2000,* edited by National Research Council, Ocean Studies Board pp. 165-71. Washington, DC: National Academy Press.

Zwally, H. Jay, et al. (2002). "Surface Melt-Induced Acceleration of Greenland Ice-Sheet Flow." *Science* 297: 218-22.

Index